T0300241

Routledge Revivals

Natural Gas Markets after Deregulation

Originally published in 1983, Broadman and Montgomery present an agenda for further research into deregulated natural gas markets by relating natural gas production, transmission and distribution with the economic function of contracts and local distribution companies. This work raises fundamental issues that could arise with the deregulation of the natural gas industry and outlines analytical methods that could be used to predict any problems that might arise and possible changes to policy. This title is of interest to students of Environmental Studies and professionals.

Natural Gas Markets after Deregulation

Methods of Analysis and Research Needs

Harry C. Broadman and W. David Montgomery

First published in 1983
by Resources for the Future, Inc

This edition first published in 2016 by Routledge
2 Park Square, Milton Park, Abingdon, Oxon, OX14 4RN
and by Routledge
711 Third Avenue, New York, NY 10017

Routledge is an imprint of the Taylor & Francis Group, an informa business

Publisher's Note
The publisher has gone to great lengths to ensure the quality of this reprint but points out that some imperfections in the original copies may be apparent.

Disclaimer
The publisher has made every effort to trace copyright holders and welcomes correspondence from those they have been unable to contact.

A Library of Congress record exists under LC control number: 83042907

ISBN 13: 978-1-138-95337-6 (hbk)
ISBN 13: 978-1-315-66735-5 (ebk)

NATURAL GAS MARKETS AFTER DEREGULATION

NATURAL GAS MARKETS AFTER DEREGULATION

Methods of Analysis and Research Needs

Harry G. Broadman
W. David Montgomery

with the assistance of
Mary Beth Zimmerman

RESOURCES FOR THE FUTURE / WASHINGTON, D.C.

Library of Congress Cataloging in Publication Data

Broadman, Harry G.
Natural gas markets after deregulation.

Bibliography: p.
Includes index.
1. Gas industry—United States—Price policy—Research—United States. I. Montgomery, W. David (William David), 1944– . II. Zimmerman, Mary Beth. III. Title.
HD9581.U5B76 1983 338.2'3 83-42907
ISBN 0-8018-3125-3

The cover was designed by Ruth Magann.

Distributed by The Johns Hopkins University Press, Baltimore, Maryland 21218
Manufactured in the United States of America

Published July 1983

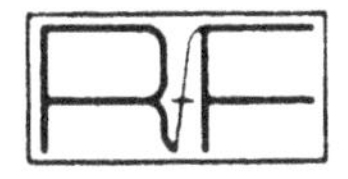

RESOURCES FOR THE FUTURE, INC.
1755 Massachusetts Avenue, N.W., Washington, D.C. 20036

Resources for the Future is a nonprofit organization for research and education in the development, conservation, and use of natural resources, including the quality of the environment. It was established in 1952 with the cooperation of the Ford Foundation. Grants for research are accepted from government and private sources only on the condition that RFF shall be solely responsible for the conduct of the research and free to make its results available to the public. Most of the work of Resources for the Future is carried out by its resident staff; part is supported by grants to universities and other nonprofit organizations. Unless otherwise stated, interpretations and conclusions in RFF publications are those of the authors; the organization takes responsibility for the selection of significant subjects for study, the competence of the researchers, and their freedom of inquiry.

This research paper is a product of RFF's Center for Energy Policy Research, Milton Russell, director. Research Papers are intended to provide prompt distribution of research having a narrower focus or a greater technical nature than RFF books.

Contents

Foreword

Seemingly irreversible shifts in the regulation of natural gas at the wellhead have altered these markets in ways that are not yet well understood. By providing conditions that allow average prices to rise to market-clearing levels at the burner tip, the Natural Gas Policy Act of 1978 changed the decision framework through which gas is bought in the field, transported, stored, distributed, and sold to final consumers. Many of the institutions and decision rules that lay beneath the operation and regulation of the industry have lost their applicability. New ways of thinking about the performance of natural gas markets are needed, and different business and regulatory practices must be developed and implemented.

Natural Gas Markets After Deregulation identifies causes of the changes in store for the downstream (transmission and distribution) segments of the industry. It provides insights into how downstream markets may adjust in this new environment, and most important, it outlines the research required to understand the new situation and to tailor appropriate private and public responses to it.

The book is organized around the major types of changes that can be expected. The authors synthesize elements of economic theories of industrial organization, information and uncertainty, and regulation to determine where they may be applied, where more data are needed, and where the theories themselves must be extended if natural gas markets are to be understood better.

This research on natural gas markets complements Resources for the Future work on other energy sources, their uses, and their regulation. It also continues the tradition of efforts to mold research agendas in energy and natural resources. While original research is its primary activity, RFF also provides an opportunity for emerging issues to be brought to the attention of the research and policy communities.

Basic support for the research on this book was provided through a contract with the Oak Ridge National Laboratory that relied upon funding from the Energy Information Administration. The resulting report met the needs of a relatively narrow constituency for an analysis of specific research needs. In recognition of the broader purpose that could be served, unrestricted RFF funds were used to expand that report, and prepare it for publication to reach a wider audience.

Research on natural gas markets offers a feast for analysts. It has few equals in terms of economic importance of the industry, immediacy of policy applications, intellectual challenge, and opportunity to advance understanding of market processes. The authors of this work supply analysts and policy makers with a menu of this rich fare.

Milton Russell
Director, Center for Energy Policy Research

Preface

This study examines changes that may occur in the downstream natural gas transmission and distribution markets as a result of field price deregulation, with a view to identifying upcoming policy issues and the research required to address those issues in an informed manner. We divide potential changes into four groups: changes in the competitive structure of downstream markets, changes in the way transactions between upstream and downstream markets are carried out, changes in pricing behavior and input and output choices of pipelines and distribution companies which remain under utility-type regulation, and changes in the downstream regulatory system itself.

In order to construct methods for analyzing these changes, we develop applications to natural gas markets from the relevant economic literature. In the case of market structural changes, this is the literature on industrial organization, in which several studies of competition in natural gas markets appear. To analyze changes in transactional arrangements, we draw on the new and growing literature on the economics of information and uncertainty, particularly that which applies to the comparison between market and non-market means of resource allocation. A narrower category of this literature, exploring principal-agent relations and the nature of optimal incentive contracting, is useful for more detailed inquiry into long-term contracting for natural gas. There is a vast literature on the economics of regulation, in which there are remarkably few applications to the downstream natural gas industry. All the major applications are at least mentioned in this study, along with a synthesis of more general studies. This literature provides a number of approaches to analysis of changes in behavior of regulated pipelines and distribution companies. Analysis of changes that may appear in the downstream regulatory regime itself requires a synthesis of all these approaches.

In each case, the review of the literature is used for two purposes: first, to see what statements can *now* be made about the direction and desirability of downstream changes resulting from field price deregulation, and second, to identify areas in which further theoretical or empirical research is required *before* definite conclusions can be reached.

In carrying out this program, we have concentrated on describing analytical approaches and general theoretical results, at the expense of providing a detailed institutional description of the history, current condition, or likely future of natural gas markets. Capable factual surveys of natural gas markets already exist, and the purpose of this study is to delineate new techniques that could be used to analyze those markets. Our description of natural gas markets is accordingly confined to the minimum required to establish points of contact with the economics literature that we review.

What emerges from this review is a systematic approach to analysis of downstream markets that involves working back from studies of final demand for gas through each of the transactions which transfer gas from the field to the burner tip. Each transaction must be studied in light of its particular set of structural, transactional, and regulatory features.

Although systematic, this approach is not simple or tightly integrated. To understand natural gas markets, a whole analytical tool kit is required. And the tools that are most useful often come from areas of research that have not as yet been subsumed as part of one grand general equilibrium theory. We have tried to record all of the major ideas and insights that we have gleaned, rather than striving for an artificial simplicity or uniformity. We have not laid out a detailed plan for developing methodology, gathering data, and reaching conclusions in any of the research areas we identify, tolerating a sacrifice of depth in order to make possible a reconnaissance of all the territory in which we thought some useful ideas might lurk.

Research for this study was supported by a contract from Oak Ridge National Laboratory and the Energy Information Administration.

Mary Beth Zimmerman served as our research assistant throughout the project. She assisted in the literature review, wrote portions of chapters 2 and 5, and checked the remainder for accuracy and consistency.

Many other people have also helped us in this study. Conversations with Cathy Abbott, Robert Means, Richard O'Neill, and Richard Wilson were particularly important in shaping our understanding of facts, issues, and what might be useful topics of study. Helpful comments on an earlier draft of this study were also provided by Paul Dickens, Robert Shelton, and Alan Tawshunsky. Glenn Gordon, George Hall, Michael Toman, and Milton Russell gave generously of their time to review the study thoroughly, and suggested major improvements. Michael Coda and Elizabeth Davis proofread and verified the manuscript. Linda Walker coordinated its typing through several drafts and has been a constant source of encouragement, sympathy, and good humor. Pamela Dyson typed the first draft, and Elisabeth Hale and ParLena Johnson assisted with typing and art work. Ruth Haas edited the manuscript. The usual disclaimer applies, that the authors made all the mistakes and those mentioned above did their best to prevent them.

NATURAL GAS MARKETS AFTER DEREGULATION

1

Introduction

The 1980s may see fundamental changes in the way natural gas is bought and sold. Either partial or complete deregulation at the wellhead will raise prices and introduce a type of uncertainty never before experienced by the industry. Business and regulatory practices that developed in a time of controlled prices and eager markets are likely to be unsuited to these new conditions.

In this volume we do not attempt, except in a most preliminary way, to reach conclusions on substantive questions of how natural gas markets will evolve in an environment of freer field prices. Rather, we suggest areas of research that would increase the understanding of market changes and the ability to deal with problems that may arise. We describe, in some detail, analytical methods that are useful for studying certain features of downstream gas markets, examine the application of these methods, and recommend specific topics for future research.

Problems of a Deregulated Market

Until 1978, producers of natural gas faced strict price ceilings on all gas sold in interstate commerce. During the 1970s these ceilings were set at such a low level, relative to the cost of additional supplies, that a chronic condition of excess demand in interstate markets was created.[1] Various regulatory measures, including moratoria on hookups to new customers, were used to allocate available supplies during this shortage.

The Natural Gas Policy Act of 1978 (NGPA) promised relief from these regulatory constraints. Although NGPA provides for only partial deregulation of natural gas prices at the wellhead, it should be sufficient to free the market from conditions of constrained supply and excess demand. Enough gas is to be released from price ceilings for wellhead prices to move to levels that could clear the interstate market. But between the wellhead and the point at which gas is consumed lies a complex network of transactions among producers, interstate pipelines, local distributors, and ultimate consumers. The prospective benefits of field price deregulation are clouded by questions about

1. Intrastate markets and pipelines have a very different history and face different current conditions. This study deals exclusively with interstate markets.

how well those "downstream" markets will adapt to freer wellhead prices.

Such questions came to the fore in 1982, as the debate on natural gas policy shifted away from wellhead pricing issues toward such "downstream" issues as the behavior of regulated pipelines and contracting practices in the industry. This shift occurred in large part because of concerns that wellhead price increases were being magnified on their way to the ultimate consumer by practices in downstream markets. In particular, provisions of long-term contracts between producers and pipelines and regulatory procedures that allowed pipelines automatically to pass cost increases through to consumers were blamed for driving delivered prices of natural gas above market-clearing levels. Pipelines were alleged to be paying more for gas purchased from production affiliates than they paid in similar circumstances to independent producers. The partial nature of deregulation under NGPA also contributed to market disorder, creating large differences in purchased gas costs among interstate pipelines and, as a result, among regions of the country. These changes in natural gas prices were not only painful to the individuals and regions experiencing them, but also disruptive to the producers, pipelines, and distribution companies who served them. Prices that deviated from uniform market-clearing levels could also be expected to cause overall inefficiencies in resource allocation.

Two bills addressing these problems were introduced in Congress during March 1983. The Reagan administration introduced a comprehensive natural gas bill pointing toward ultimate deregulation of all natural gas prices at the wellhead, and Representative Gephardt introduced an alternative bill reimposing controls on all natural gas prices. Both bills also deal at length with contracting practices and downstream regulation.

Need for Research

The current situation in natural gas markets and the legislation before Congress demand attention to questions of how downstream markets will adapt to freer wellhead prices. Whether or not natural gas legislation is passed in 1983, these questions are likely to remain. The problems on which policy debates now focus are manifestations of deeper characteristics of natural gas markets. These are largely unstudied, and as a result there is little analytical basis for predicting likely directions of change and assessing the need for policy intervention in downstream markets.

Current questions about pipeline transactions with production affiliates and pipeline market power concern structural features of the downstream market. These may change over time and in response to deregulation, but little is known about why and how changes will occur. For example, at some points in the system there are few sellers (or buyers), so that adequate competition may not exist.

Problems of contract terms arise because most downstream transactions involve long-term contracts that may be ill suited to the conditions created by field price deregulation. To a minor extent, production and transportation of natural gas are conducted as integrated activities, and some gas is sold on a "spot" basis. But for the most part, pipelines purchase gas under long-term contracts which allow neither the price nor the quantity of gas to be adjusted in response to market conditions. Whether such problems will persist depends on how transactional arrangements evolve. It is conceivable that contracting practices are an artifact created by field price regulation, and that the problems will disappear as contracts are renegotiated. It is also possible that underlying physical and technical features of the natural gas industry make a nonmarket allocation process, such as that found in an integrated firm or under a long-term contract, superior in efficiency to a "spot" market system in which prices and quantities are redetermined frequently and buyers and sellers freely change partners. If this occurs, contracting practices may change, along with the balance among integrated, contractual, and spot transactions. Research is badly needed to determine what these changes will and should be.

Concerns about the motivation and behavior of regulated pipelines also point to the pervasive influence of downstream regulation. Under NGPA, interstate pipelines remain under the jurisdiction of the Federal Energy Regulatory Commission (FERC), and local distribution utilities are regulated by agencies of state governments. These regulatory agencies influence the behavior of pipelines in bidding for gas supplies, investment decisions involving capacity and storage levels, allocation of rising gas costs among customers and uses, and the reliability of service. It is impossible to understand in what way and how efficiently downstream markets will respond to gas price increases without studying their regulatory environment.

Careful examination of the structure of the natural gas industry, of how its transactions are carried out, and of how it is regulated, can provide a basis for discerning the deeper issues that lie behind current problems and policy concerns. Such analysis can also aid in understanding what actions might improve the efficiency of gas markets. Some problems that now attract great attention may be ameliorated by pending legislation and others may solve themselves. Some are likely to persist and new problems may appear—either as results of short-sighted legislation or because of the internal logic of downstream markets.

Thus research on natural gas markets after deregulation must serve two functions: it should predict problems that will persist or appear afresh in the years ahead and it should provide an analytical basis for designing policies to cope with those problems.

Plan of the Study

This study is organized around the structural, transactional, and regulatory features of natural gas markets. Each of these separate but related features must be analyzed in order to develop some sense of how natural gas markets will evolve. Each may depart from accepted conditions for efficient allocation of economic resources.

Study of these features of downstream gas markets requires new and sometimes sophisticated analytical methods. Thus much of this volume is devoted to describing new currents of thought in the literature on industrial organization, price theory, and the economics of public utilities and regulation. We suggest ways in which these methods can be used to analyze problems in downstream markets and, in the course of describing those applications, achieve some insights into emerging issues.

These applications show that it is possible to bring general propositions of economic theory to bear on the particular problems of natural gas markets. Some preliminary insights are generated, but for the most part we find that further research, either on a theoretical or empirical level, is required before definite conclusions can be reached about potential changes in downstream markets.

Underlying the economic features of downstream markets are a number of technical characteristics of natural gas production, transmission, and distribution. These are described in chapter 2 and related to economic features of the industry. Chapter 2 also provides a brief overview of changes affecting downstream markets that can be expected to result from field price deregulation, and discusses some of the problems that may arise.

The subsequent chapters deal with the economic features of the industry; chapter 3 discusses structural conditions that may give rise to market power at some points in the system; chapter 4 addresses the question of how transactions are conducted in natural gas markets; and chapter 5 analyzes regulation of pipelines and local distribution companies.

Each chapter surveys a separate branch of economics, in which new developments provide potentially fruitful methods for analyzing a particular aspect of downstream natural gas markets. Different analytical methods are required for understanding different features of these markets. They include studies of market structure, analysis of nonmarket allocation processes, and the economics of regulation. Because natural gas markets have not been studied extensively with these methods, general developments in economic theory and their applications to industries that have problems and features similar to natural gas are emphasized. Each chapter introduces the issues to be addressed, describes the available methods of analysis, and illustrates how they can be applied to the particular set of natural gas issues.

Chapter 6 deals with two issues of regulatory policy that are not directly addressed by current legislation and which require all the analytical methods discussed previously. These are the role and economic function of interstate pipelines in a market with freer wellhead prices, and regulatory allocation of costs between jurisdictional and nonjurisdictional customers. This analysis provides an example of the type of question about downstream markets that is likely to persist well into the eighties, and which the research described in this study could help resolve.

The final chapter draws together the discussions of likely market developments, literature reviews, and research recommendations. It highlights the principal areas of concern about downstream adaptation to field price deregulation and the most productive areas of future research.

2

The Natural Gas Industry in a Deregulated Environment: An Overview of Policy Issues

In order to introduce some important factors about the industry which shape both its current problems and the strategy for future research, it is helpful to sketch out a description of natural gas markets. This chapter provides some background on characteristics of natural gas markets and on the policy issues and intellectual puzzles that analysis of those markets could address.

From Field to Burner Tip

The Physical System[1]

Natural gas is a hydrocarbon, mostly methane (CH_4), which is found by drilling into an underground reservoir formed when gas is trapped beneath a layer of impermeable rock. Natural gas is found either by itself or in association with oil. The major producing areas in the United States are located in Texas and Louisiana and include offshore fields in the Gulf of Mexico. In 1980 these two states produced 14 trillion cubic feet (tcf) out of the total 21 tcf produced in the entire United States. Only four other states produced significantly more gas than was consumed within their borders: Oklahoma and New Mexico together produced 3 tcf, Kansas produced 0.7 tcf, and Wyoming produced 0.4 tcf (AGA, 1982). Some gas is also produced in the Appalachian region that extends from the Gulf Coast up through Ohio and Pennsylvania, and smaller amounts are produced in Michigan, the Rocky Mountain states, and California. A network of long-distance pipelines connects these producing regions with end-use markets throughout the United States.

Gas from all wells located in a given "field" (a reservoir or set of reservoirs associated with a particular geological structure) is collected through small diameter pipes ("gathering lines") and delivered to field processing facilities. These facilities, termed "lease separators," separate natural gas from the oil with which it is associated and remove some other heavy liquids from the gas stream. The gas is then transported by pipeline to a processing plant, where remaining natural gas liquids (mostly ethane, propane, and butane) are removed. The resulting dry gas

1. A more detailed description of the process of producing and transporting natural gas, on which this section is based, may be found in EIA (1982c).

stream is ready for long-distance transportation through "trunk" pipelines.

Pipelines are interconnected in ways that allow gas to be moved from one trunk system to another, and to be shifted among regions in response to changes in demand. Transactions between pipelines are common, and are referred to as "off-system" sales.

Long-distance pipelines are constructed of welded steel pipe, up to 42 inches in diameter, and operate at high pressure, which is controlled by a series of compressors at intervals along the pipeline. Typically fueled by natural gas taken from the pipeline, these compressors provide a major component of the variable cost of transporting natural gas because the cost of compressor fuel depends on the price of natural gas. Increasing or decreasing the pressure in a pipeline provides limited but important opportunities for matching fluctuating demand and supply. Building up pressure to meet an expected transitory increase in demand is called "packing the lines"; letting the pressure drop allows demand to increase for a short time relative to production.

Another important technical aspect of the transmission and distribution system is its use of storage. Demand for natural gas is highly seasonal, peaking during the winter when space heating demand is at its highest. It is more costly to build sufficient production and pipeline capacity to meet this demand with gas flowing through the system than to construct storage at a point between the field and the burner tip. Depleted gas fields located near points of use are often converted to storage units. When storage is available, gas can be produced at a nearly constant rate throughout the year and transported to a storage site near its point of use. This makes it possible to economize both on production capacity and on the transmission capacity of the pipeline, which can be sized in light of the average quantity of gas to be delivered throughout the year rather than the peak quantity demanded. Gas is injected into storage during the summer season when demand in cyclically low and then withdrawn during the winter to meet peak demand.

Between September 1981 and August 1982 interstate pipeline monthly sales varied from a peak of 1.7 tcf in January 1982 to a low of 0.94 tcf in August 1982, a peak-to-off-peak demand ratio of almost two to one (EIA, 1982c). Total gas in storage owned by interstate operators was highest in October 1981 (4.7 tcf) and lowest in March (3.5 tcf). Of the gas in storage, a large fraction (about 2.4 tcf) is "base gas," which maintains pressure in the storage field and cannot be removed under normal circumstances.

Interstate pipelines own most storage facilities, though some are owned by large distribution companies. The former deliver gas to large industrial customers and electric utilities (direct "mainline" sales) located throughout their system and to local distribution companies ("city gate" transactions). These in turn deliver gas to final consumers through a system of low-pressure pipes.

Natural gas consumption has a much more even geographic distribution than does gas production. Texas, California, and Louisiana are the states with the largest gas consumption (3.4, 1.7, and 1.5 tcf, respectively in 1980), followed by Illinois (1.1 tcf), Ohio (0.9 tcf), Michigan (0.9 tcf), Pennsylvania (0.7 tcf), and New York (0.7 tcf) (AGA, 1982). The principal uses of natural gas are for space heating, for process heat and boiler fuel in manufacturing industries, and as a chemical feedstock. In the intrastate market of the Southwest, natural gas has also been used as a boiler fuel by electric utilities. In 1981 residential and commercial gas consumption (primarily for space heating) was 6.9 tcf. Industrial consumption was 7.1 tcf, of which 0.8 tcf was from direct mainline sales to industrial customers by interstate pipelines. Electric utilities consumed 3.6 tcf, of which 0.6 tcf was direct mainline sales. Pipeline fuel was 0.6 tcf (EIA, 1983).

Economic Features

To bring a particular gas field into production, it is necessary to drill and equip gas wells, to construct gathering lines and lease separators; and if the field is isolated, to build a system of pipes which will transport the gas to a location where it can enter a processing plant or major pipeline system. Once it is completed, the output of the gas well can be varied from zero, by capping or shutting in the well, to a rated maximum production which is determined by well size, spacing, and characteristics of the reservoir. It is desirable to avoid operation below rated capacity because of the high fixed costs that must then be recovered on a smaller quantity of production. Moreover, the common property problem applies to gas production because many different operators may drill wells into the same underground structure. Thus if one well is shut in, its owners may find that gas which they expected to produce is being drained by other operators. Because of this drainage problem, both regulatory and contractual provisions are written to ensure more or less uniform extraction of gas by all operators in a field.

The costs of drilling, equipping, and connecting a well to a pipeline system are largely fixed investments which must be made before gas can be brought into production. Actual costs of operating the gas well for most of its life are relatively small in comparison. However, periodic maintenance and workovers (rehabilitation of wells to restore production rates) are required to keep wells in operation and additional drilling to increase flow rates is always possible. Thus the rate at which gas can be produced from a field depends on prior investment decisions and on current operating practices.

In a typical transaction, a natural gas producer will explore for, find, and develop a field. Once the existence of gas has been proved, and the characteristics of a reservoir determined, an estimate of the reserves involved in the particular geographic location is made. Then the producer will arrange for a natural gas pipeline to buy this gas. The pipeline in turn transports the gas and resells it to another customer downstream. In essence, the transaction between a pipeline and a producer involves selling rights to whatever reserves exist under a particular tract of land. Interstate pipelines have generally operated as "private carriers," transporting gas which they own. But at times a downstream party will purchase gas directly from a producer, and a pipeline will serve as a "contract carrier" transporting that gas for an agreed-upon fee.

Most transactions between producers and pipelines take place under long-term contracts that specify prices and quantities of gas to be delivered. These contracts serve to provide performance incentives and to allocate risks among their signers. Typical terms in contracts between producers and pipelines include price escalators, specifying how future prices will be determined, "take or pay" provisions specifying a minimum fraction of a field's potential output that a pipeline must pay for whether or not delivery is taken, various buyer protection provisions, grace periods during which take or pay obligations may be made up or prices and delivery terms renegotiated, and provisions requiring pipelines to transport to others gas that they do not take (EIA, 1982a).

Although most natural gas is produced by an independent producer and sold under contract to an unaffiliated pipeline, there is also an integrated mode of operation. Approximately 12 percent of interstate gas is produced by production affiliates of pipelines and sold to their parent company (Portman, 1982). Gas produced by pipeline affiliates is also sold by those pipelines to others rather than retained for use in their own system. Integration between pipelines and producers has been more prevalent in the intrastate than in the interstate market (Mulholland, 1979), but an increasing degree of backward integration by interstate pipelines is reported (Cowan and Hagar, 1982). Spot purchases involving sale of a particular quantity of gas for prompt delivery are another way to arrange transactions between producers and pipelines. Such transactions are reported to be more prevalent in the intrastate markets than the interstate. In short, while long-term contracting is the rule rather than the exception in the interstate markets, it seems less important in the intrastate market, which until NGPA, was free of field price regulation.

Interstate natural gas pipelines are regulated by the Federal Energy Regulatory Commission (FERC), which has jurisdiction over interstate pipeline sales to retail gas distribution companies and over "off-system" sales to other pipelines and to retail gas distribution companies. The FERC regulates pipelines by approving tariffs that will provide a "fair rate of return" to a pipeline's capital investment. Direct mainline sales to industrial customers are not under the jurisdiction of FERC, but they are influenced by regulation because portions of a pipeline's costs are common to its jurisdictional and nonjurisdictional sales. The FERC exercises control over the allocation of joint costs of pipeline capacity and, under the incremental pricing provisions of NGPA, also directs the allocation of variable (purchased gas) costs. In addition, the FERC must provide a "certificate of public convenience and necessity" before construction of any new natural gas pipeline or any other related facility is allowed. The FERC also has jurisdiction over abandonment, which is the termination of a contract either to purchase gas or to sell gas by a pipeline.

The process of FERC regulation is typical of commission regulation of public utilities. At periodic rate proceedings, a pipeline is required to forecast throughput and, based on this forecast, to submit a schedule of prices which would provide sufficient revenues to cover costs and a "fair" return on assets (the rate base). Hearings cover such subjects as the reasonableness of the sales forecast, the proper valuation of the rate base, the fair rate of return, and the rate structure for different categories of customers and use. Pipelines are also allowed, after notification but without a hearing, to pass on to their customers any increase in the cost of purchasing natural gas. This "purchased gas adjustment mechanism" eases the regulatory burden that would be created if a new rate review were required for every increase in gas cost. It also reduces the lag between the time when increased costs are incurred and when they are recouped from customers.

Retail distributors who purchase gas from interstate pipelines in turn resell it within a local service area. They are regulated by state public service or public utilities commissions. Based on the costs they incur in purchasing gas and their own capital investment, these distributors determine the prices paid by and the availability of gas to residential, commercial, and industrial customers, including electric utilities within their service areas. Their relations with interstate pipelines typically take the form of long-term contracts in which they often agree to "minimum bill" provisions, obliging them to take at least a minimum quantity of gas from a pipeline at prices based on the pipeline's costs. Some agree that their full gas requirements will be purchased from the contracting pipeline; others that all sales within a geographic region will come from the subject pipeline. Distribution companies, on the other hand, often sell gas on a noncontractual basis, for example, to residential customers who face no requirement as to the minimum amount of gas they must use.

Since it is never possible, given the vagaries of weather and economic conditions, to ensure that all gas demand will be met, regulators also set curtailment priorities. Both interstate and local distribution companies are required to cut off customers in a particular order. If sufficient gas is not available to meet all demand, those customers with a capacity to burn a substitute fuel are among the first to be cut off. These usually include large industrial users who have boilers that can be fired either with fuel oil or with natural gas. Extreme measures are taken to prevent the shutoff of gas to residences and small businesses because of the danger involved. If gas supply to a residence is interrupted, pilot lights go out and when service is restored, an explosion can occur. Thus an expensive process of checking every gas-using appliance would be necessary before residential service could be restored.

From Shortage to Glut

The NGPA was enacted after a half decade of natural gas shortages.[2] These had been particularly acute in states lacking significant intrastate production; interstate price controls discouraged producers from releasing supplies to these markets and several unusually cold winters aggra-

2. For a brief survey of issues in natural gas regulation prior to passage of the NGPA, see Brauetigam (1981).

vated the problem. By raising price ceilings and allowing an increasing portion of gas reserves to be sold at market-determined levels, and by correcting the apparent imbalances between inter- and intrastate markets, it was hoped that NGPA would alleviate the supply shortfalls while limiting the transfer of inframarginal rents ("windfall profits") to gas producers.

By 1982, it was apparent that the goal of eliminating supply shortfalls had been reached. Some producers were willing to supply more gas than they could sell. Some interstate pipelines found themselves caught between downstream markets which could not absorb sufficient quantities of gas at contracted prices and contract provisions with field producers which stipulated minimum purchases at predetermined prices. Yet, while apparent surpluses grew, end-use consumers protested prices which, by the last quarter of 1982, were as much as 24 percent higher than a year earlier. By the end of 1982, many observers held that the average price of natural gas met or exceeded its market-clearing level (EIA, 1982c).

This cycle from shortage to glut has prompted calls for reform of end-use pricing, contracting practices, and the regulation of interstate pipelines. As background for understanding these issues, the remainder of this section reviews the principal provisions of NGPA and the market conditions currently faced by pipelines and residential and industrial customers.[3] The final section discusses the continuing issues and problems that may appear in downstream markets.

The Natural Gas Policy Act

Until 1978, the price at which gas was sold by a producer to a pipeline engaged in interstate transportation was subject to federal price ceilings. Gas sold to an intrastate pipeline was free of ceilings. The Natural Gas Policy Act (NGPA) of 1978 applied price controls to all gas, regardless of its status in interstate or intrastate commerce, but provided also for a gradual lifting of price ceilings for some categories of gas.

3. Further discussion of the issues raised by NGPA may be found in Means (1982) and Russell (1982).

The NGPA identified several categories of "old gas," interstate gas from fields that were in operation prior to 1977, priced primarily on the basis of when the well was brought on line. Scheduled price escalations were designed to keep real gas prices near their original levels. Section 104 gas includes all old gas under contracts prior to NGPA enactment, and section 106 covers "rollover contracts" on previously producing wells. By the end of 1982, old gas accounted for some 56 percent of total gas sales and the ceiling price ranged from 45.5 cents to $2.24 per thousand cubic feet (mcf).[4] In 1979, these prices were 35.9 cents to $1.77/mcf.

New gas prices were set somewhat higher than those for old gas to encourage the development of new reserves and marketed supply. Section 103 gas is comprised of gas from new sources which are most easily developed and prices are allowed to rise with the rate of inflation. Section 102 gas is from more expensive sources, such as the outer continental shelf and wells some distance from existing fields; prices for these sales were initially the same as for 103 gas, but have been allowed to increase somewhat faster than the inflation rate. By 1982, the price for section 102 gas was about 21 percent higher than for 103 gas. Sales for both categories have increased significantly over the last five years. In addition, a category of stripper wells (section 108) was created which qualifies for a slightly higher rate ($3.51 as of December 1982).

One type of "new gas" deserves special attention. The NGPA provides for certain types of expensive production to qualify for unregulated prices. This section 107 "high cost" or "deep" gas has been the subject of a great deal of controversy because some contracted prices reached a high of around $9.00/mcf in early 1982. The actual range of high-cost gas prices in mid-1982 was $3.15 to $9.65/mcf (EIA, 1982b). While only about 7 percent of all gas marketed in 1982 was deep gas, the large differential between these prices and those for other categories means that

4. For a detailed breakdown of natural gas production and price statistics, see the most recent issues of the Energy Information Administration's *Natural Gas Monthly*. Figures reported are contianed in the December 1982 issue.

average pipeline acquisitions costs will be significantly influenced by high-cost gas purchase decisions.

In addition to these domestic supplies, about 5 percent of natural gas consumed in the United States is imported, principally from Canada and Mexico, with lesser amounts of liquid natural gas contracted for from Algeria. The 1982 Canadian and Mexican price of $4.95/mcf was set by negotiations between the two governments and price changes are not directly controllable by legislative action.[5] Algerian sales are subject to (pending) FERC approval.

In addition to establishing price categories for interstate natural gas wellhead sales, the NGPA controls most intrastate sales and specifies the regulatory authority to implement the act. Oversight of gas sales is conferred upon the FERC, but the actual degree of control which FERC has over individual contracts is a point of contention. Under its "incremental pricing" provision, the NGPA also affects end-use pricing by requiring that pipeline expenditures on high-cost gas be passed through separately to certain industrial customers, until their prices reach an "alternative fuel price."

Wellhead Purchases

The average wellhead acquisition cost of gas for some pipeline companies is more than double that of other companies. As of mid-1982, average pipeline purchase prices ranged from $1.34 to $2.90/mcf; individual purchases within each company's portfolio are often much smaller or larger (EIA, 1982b).

Average cost differentials exist because the mix of gas purchased from the NGPA price categories varies across pipelines. Some pipelines were able to contract for substantially larger quantities of old, inexpensive gas than were others, and are referred to as having a larger "price cushion." Pipelines do not absorb their contracted reserves at equal rates, so that some pipelines have a larger or smaller proportion of old gas in their recent purchases than in their inventory of reserves. As a result of these two factors, recent takes of old gas have varied across pipelines from a high of nearly 80 percent of total purchases to a low of about 30 percent (EIA, 1982b).

While the ratio of old to new gas is an important determinant of overall cost differentials, it does not take into account the large price disparity among new and deep gas categories. Most pipelines found themselves in need of acquiring supplemental—and expensive—reserves after the shortfalls of the mid-1970s. The extent to which they purchased deep gas rather than less expensive new gas is presumably a function of geography, previous supplies, resale marketability, and relative bargaining positions vis-a-vis producers. High-cost gas purchases by pipelines range from zero to almost 15 percent of total takes (EIA, 1982b).

Finally, within each category there is a significant range of contracted prices. This is most evident in high-cost gas sales, but is true for each of the other categories as well. The average price for old gas (104 and 106), for example, varied from 68 cents to $1.78 /mcf in mid-1982 (EIA, 1982b). This price variation may reflect not only price ceilings but also the differential bargaining power of pipelines and differing long-term marketing strategies. The price differentials may also partially reflect differences in nonprice contractual obligations; pipelines may "trade" higher prices for more favorable terms.

The ratio of old to new reserve holdings is decreasing for all pipelines since old gas, by definition, cannot be replaced. Even without further changes in the ceiling or contracted prices for any category, average wellhead prices will rise as old gas is depleted. Because of differences in the holdings of old gas reserves and in the rate of purchases, however, this price increase differs among pipelines.

The failure to take quantities of natural gas specified in supply contracts can subject pipelines to any of a variety of penalties. A survey of contract terms (EIA, 1982a) has shown that older contracts, involving gas with relatively low prices, are less stringent in their requirements or penalties than newer contracts for higher cost gas. As a result, when demand falls,

5. In April 1983 the Canadian government announced a $.50 reduction in this price.

pipelines may find it advantageous to reduce their purchases of old, low-priced gas while continuing to take delivery of high-priced gas. These purchase patterns can create an apparent supply glut, with some producers willing to sell more gas than is being taken at contracted prices. The pattern of gas production that results may also fail to meet given demand at minimum cost, as gas with higher production costs may be taken while that with lower production costs is shut in.[6]

The combination of fixed gas categories and the price cushion enjoyed by some interstate pipelines suggests the possibility that pipelines enjoying low average costs might be able to "bid up" the price of deregulated gas to levels far in excess of market-clearing prices, thereby raising the national average cost of gas. Contracts for $7.00 and $8.00/mcf seem to bear out this development. Yet, while deep gas prices are certainly high, it appears that large cushion pipelines are not principally responsible for these purchases (EIA, 1982b). A large cushion is generally associated with a large reserve-to-production ratio since older reserves with low ceiling or contracted prices have tended to have higher reserve-to-production ratios than do newer reserves. The pipelines which are most aggressively contracting for high cost gas seem to be those lacking substantial reserves and these generally have been low cushion pipelines. Thus pipelines which have been willing to pay these costs may be paying for a more secure supply stream with the higher than average price.

To the extent that partial decontrol permits some gas to be bid above the market-clearing level, NGPA price regulation has the effect of transferring producer surplus from one set of producers (those without much high-cost gas, generally large oil companies) to another (those who principally produce high-cost gas, generally smaller, independent producers). As a result, total deregulation after 1985 might involve average prices to consumers not much different than prices under partial deregulation (DOE/OPPA, 1981).

Residential and Industrial Rates

Pipeline wellhead purchase practices and differential pipeline costs are reflected in the bills of end-use customers, but the relationship between residential and industrial rates and wellhead prices is not generally that straightforward. Rate structures established by state regulatory commissions and the possibility of direct industrial sales means that costs are not necessarily passed through evenly to all classes of customers.

Differences in residential natural gas bills result in part from the average cost differentials of the pipelines discussed above and in part from different market conditions and transportation costs. In the continental United States, average residential rates for natural gas varied from $45.24 per hundred therms[7] in Miami to $80.27 in Boston, with a national average of $59.58 in October 1982 (EIA, 1982c). In general, the Northeast tends to have the highest average residential rates, followed closely by the West Coast. Undoubtedly, a significant portion of this range is attributable to differential transportation costs. The U.S. city average residential rate increased some 24 percent from October 1981 to October 1982, and substantial additional purchased gas adjustment (PGA) pass-throughs were made throughout the next several months.

Rates for direct industrial sales have also increased significantly. From July 1981 to July 1982, the national average rate rose 20 percent and, again, this does not include substantial winter 1982–83 increases. Rates for many customers are now as expensive as or more expensive than residual oil. In every month in 1982, sales to industrial customers were down significantly from 1981 while total gas sales declined only slightly.[8] Industrial customers are subject to price variation for reasons similar to those in the residential market, but when purchases are made

6. Since prices for various types and vintages of gas are heavily influenced by past regulation and contracts, it is not necessarily true that relative prices of natural gas reflect relative costs of production. Thus failure to take lower *priced* gas does not necessarily imply inefficient resource allocation in the form of shutting in low *cost* gas.

7. A therm is approximately equal to 0.1 mcf of gas.

8. Of course, aside from the effects of higher relative prices for gas, the recession during this period undoubtedly contributed to a decline in industrial gas consumption.

directly with the pipeline, the variety of pricing arrangements is likely to be larger (EIA, 1982c).

Pipeline companies have observed this erosion of their industrial load with some alarm. Industrial customers tend to purchase gas at a more even rate than do residential users and help the pipeline limit average costs through more efficient use of their capital stocks. Higher operating costs and a reduced number of customers to share them could result in higher average residential as well as industrial rates. Given this situation, both pipelines and industrial users have requested FERC permission for pipelines to reduce industrial rates when necessary to preserve operating load.

Both residential and industrial customers have been concerned about the causes of rate increases, particularly what portion of recent rate hikes is the result of increased fuel costs and what is the result of higher transmission costs, both from the pipeline and the local utility. The American Gas Association (AGA) recently contended that about 60 percent of the total bill is attributable to the cost of gas itself, an increase from 32 percent in 1972. The Natural Gas Supply Association (NGSA) contends that gas purchases are only about 42 percent of retail rate (see *The Oil Daily*, 1983, p. 3). The difference in estimates results from different definitions of gas (domestic or all, etc.) and of "average retail price." It seems clear, however, that recent large increases in gas prices have raised the portion of bills attributable to fuel costs. Decreased demand and more uneven loads may contribute to increasing transportation costs as well.

Contracts

Questions about contract terms, such as take or pay provisions, escalator clauses, and minimum bills ultimately concern the nature and role of long-term contracts in the natural gas industry and should be examined from that perspective. The existence and nearly universal use of long-term contracts is probably in itself a function of the long-term nature of production. The specific terms of contracts, and recent changes in them, however, may be as much responses to the prevailing price regulation as to deeper conditions.

Without price controls, natural gas markets take on some of the general characteristics of commodity markets. Demand is uncertain and fluctuates with changes in the level of economic activity and in weather and long-term regional demographic shifts. Reserves and supplies are themselves uncertain, at least at the level of individual fields. If deregulated natural gas demand is sensitive to changes in alternative fuel prices, the market will be even more uncertain.

Natural gas production has the additional quality that most wells tap into common fields. The threat of losing reserves because of drainage by neighboring wells creates an incentive for producers to insist upon maximum production rates and sales. Thus producers have been increasingly insistent upon take or pay obligations and other provisions which compensate them for failure to capture as much of the field as possible.

The long-term nature of investment, by producers and pipelines alike, is the most frequently cited motive for contracts that reduce capital risks. Each of these aspects of the industry will persist whether or not current controls on field prices remain in force. While specific provisions such as high take or pay requirements might be modified, the producer's incentive to ensure maximum flow remains, and some form of such insurance is likely to remain a part of contract negotiations.

Some contract provisions may, however, be a consequence of regulation. Price ceilings mean that pipelines anxious to outbid their competitors for reserves must find nonprice alternatives with which to lure producers into making the desired supplies available. Guarantees of adequate and consistent purchases are one way to increase the effective compensation offered for gas supplies. During most of the 1970s this proved to be a fairly cheap form of payment since general supply shortages meant that takes would rarely be reduced and that "shopping around" from supplier to supplier was unrealistic. It was also an effective way to assure customers and regulators alike that adequate suppliers were being aggressively sought.

After passage of NGPA established prospects for eventual deregulation of new gas supplies, pipelines could also compete by offering to pay high prices for future, deregulated gas. Offers for high gas prices in the future could help "pay for" increased supplies of low-priced gas in the present. These offers could be embodied in the "most favored nation" clauses or "indefinite price escalators" found in many contracts. While a few pipelines included "market out" provisions ensuring against unmarketable price escalations, these clauses would have largely eliminated incentives provided by the prospect of high future prices.[9] When deregulation makes it possible to pay market-determined prices for current production, the role of radical price escalators without escape clauses as an incentive will diminish. In this case, such provisions would not be expected to reappear in new contracts. The problems they might create should, therefore, be considered transitional.

Pipeline contracts with local distribution companies often contain a "minimum bill" clause which is similar in function to the take or pay clauses in wellhead contracts. This provision also serves to ensure pipelines against loss caused by inadequate use of capital stock and "shopping" by distributors. In addition, it makes it easier for the pipeline to meet its own take or pay obligations by discouraging significant load losses. More flexible supply contracting could partially reduce pipeline desires for minimum bill provisions, but as with take or pay provisions, it would not eliminate them.

Conclusion

The NGPA introduced a new structure of prices and an element of deregulation into an industry whose ways of buying and selling gas were conditioned by years of strict wellhead price ceilings. Some of the problems described in this section arise from the provisions of the NGPA and from contracts and regulatory practices inherited from the years of regulation. These may be resolved as the industry becomes accustomed to operating in a freer market environment, though the speed, direction, and costs of change are not yet known. But other problems arise from basic characteristics of the industry, and from regulatory practices and ways of arranging transactions that are not merely artifacts of field price regulation. The next section discusses what may be permanent features of downstream markets under field price deregulation, and some of the issues that arise as a result.

Living with Uncertainty

Short-Run Shocks

Any form of effective deregulation of field prices will cause enduring changes in the nature of uncertainty in natural gas markets (Russell, 1983). These markets have always been vulnerable to random shocks, as a result of changes in weather, levels of economic activity, or prices of alternative fuels. But for producers, pipelines, and distributors, those shocks were muffled by price ceilings and excess demand. Shifts in demand simply caused greater or lesser curtailments of customers, but all sellers of gas remained able to find ready buyers at the legal price for any gas they could (or wished to) obtain. Consumers faced considerable uncertainty in the quantity of gas they could obtain, but shocks did not translate into revenue risks for producers or pipelines. By eliminating excess demand, field price deregulation moves some risks back from the consumer to upstream segments of the industry. Regulation and contract terms determine the allocation of risk among upstream parties, and practices adopted during a time when risks facing those parties were minimal may produce perverse results in a risky world.

Along with risks, field price deregulation creates new opportunities for producers, pipelines, and distributors. No longer constrained by inadequate supplies and curtailment priorities, pipelines and distributors can seek out new markets; producers can explore and develop resources

9. A "most favored nation" clause ties the price in a given contract to the highest price (or average of the highest two or three prices) paid to any nearby producer. An "indefinite price escalator" sets the price of gas equal to some fraction (or multiple) of the price of an alternative fuel, such as distillate fuel oil. A "market out" provision allows the purchaser to reduce prices unilaterally or to renegotiate a contract if it becomes impossible to resell gas at the contracted price.

that would not have been economic under pre-NGPA price ceilings; and in all parts of the market efficient resource allocation becomes possible through matching marginal costs of supply with willingness to pay. Market-determined prices provide the incentive for these activities at the same time that they put new risks on their undertakers.

The problem of living with uncertainty is understanding the tradeoff between risk and incentives. The way in which that tradeoff is made will have much to do with the future stability or instability of gas markets. The issues can be summarized in two questions:

> Will prices, costs, and revenues return to long-run stability after a transitional period, or will natural gas markets exhibit permanently the price instability that characterizes other markets for primary commodities?
>
> Will prices adjust flexibly enough to clear gas markets continuously, or can future cycles of gluts and shortages be expected?

The pipeline and distribution segments of the natural gas industry are imperfectly competitive, capital intensive, and tightly regulated. They share these characteristics with the electric utility industry, which in the late 1960s was suddenly confronted with analogous uncertainty about demand and increases in costs. Regulators of electric utilities developed new procedures in response (Joskow, 1974), but a long period of financial stress and operation with a suboptimal generating mix followed.

The structure of these transactions and the regulators' reactions to these uncertainties will determine the future performance of natural gas markets. Some provisions of current contracts appear unsuitable, even though these provisions left all parties with stable revenues, supplies, and costs during the era of wellhead price regulation. But with prices rising in unexpected ways, the same provisions introduce costly rigidities in the market. With definite price escalators and minimum delivery requirements, excess supply can be created when burner tip demand falls because of events outside the gas market, such as lower prices for alternative fuels and reduced economic activity. When demand falls, contractual obligations and penalties can also lead pipelines to continue using costly sources of gas while reducing use of less costly ones. The result is inefficient production of the quantity of gas that *is* demanded. A producer benefits from more stable revenues if it has sufficiently stringent take or pay provisions relative to other producers, but less well protected producers, pipelines, and consumers can suffer.

Some costs of this type may be inevitable when greater uncertainty is injected into markets with the high fixed costs and potential market power of natural gas. However, changes in contract terms, in the direction of a more even distribution of risks across parties to the contract and greater pricing flexibility, seem likely to reduce costs of long-term contracting.

Regulatory practices also affect the ability of downstream markets to handle uncertainty. For example, fuel adjustment mechanisms were introduced in the electric power industry during the 1960s to avoid the Hobson's choice between frequent (and expensive) rate proceedings and regulatory lag (Joskow, 1974). They made it possible for unexpected increases in fuel prices to be reflected rapidly in consumers' bills, thus protecting the financial integrity of electric utilities as well as providing more accurate incentives for consumers' decisions on electricity use. Purchased gas adjustments serve the same functions in natural gas markets, but their operation would be severely restricted by the administration's natural gas bill. Such restrictions may reduce swings in the prices consumers see, but can also trap pipelines and distributors between rising costs and controlled prices. Under these conditions, their willingness to bid for gas supplies could be reduced, and gas shortages could again develop. Thus much as contractual provisions hinder downward adjustments of gas prices to meet the market, regulatory actions could prevent upward adjustments to cover cost. Taken together, these constraints could exacerbate the instability of natural gas markets by preventing the price movements that would signal the need for short-run increases or decreases in supply.

At a more subtle level, the design of contracts between producers and pipelines and the terms of sale by pipelines and distributors to firm and interruptible customers can increase or decrease market stability. A necessary condition for efficient resource allocation is matching customers' willingness to pay for secure supply with the costs of investing in requisite production, storage, and transportation capacity to meet peak demand.

Long-Run Uncertainty

Natural gas markets are subject to long-run uncertainty, as well as to random shocks caused by shifts in demand or costs. Perhaps the most important long-run question is the extent to which gas will continue to be used by industry. Uncertainty about the long-run cost of gas supplies and about the determinants of industrial fuel use lead to wide variation in estimates. But institutions in the gas market may deal well or poorly with this uncertainty. Whatever the future holds in terms of supply costs and industrial demand, regulations regarding cost allocation are likely to have a strong influence on the viability of the industrial market. If regulators try to protect residential markets from rising prices by allocating additional costs to industrial sales, the industrial market may be lost prematurely or unnecessarily. On the other hand, competition by regulated pipelines for unregulated industrial sales will lead to underpricing in those competitive markets, moving costs into and increasing prices on jurisdictional sales. The reliability of service that can be provided to industrial customers may also affect their demand. This reliability will be affected both by the arrangements pipelines and distributors make with upstream suppliers and by the regulatory treatment of sales to lower priority customers. Innovative rate designs could also aid in achieving economically efficient decisions about whether or not gas should be used in the industrial market.

Public Policy Issues

The changes appearing in natural gas markets raise questions that should concern regulators, other policymakers, and participants in the market. These include questions of how and whether downstream markets should be regulated, whether and by whom actions should be taken to make transactions more responsive to market conditions, and what scope there will be for competition as a disciplinary force in the new market. Many of these concerns come together in what may be the central issue of the years ahead: the role of pipelines in a freer and more uncertain market.

Federal and state regulatory agencies with jurisdiction over downstream markets face thorny questions of how to deal with the new cost conditions and uncertainty appearing in gas markets. Four responses by regulated pipelines and distributors to higher prices and new market opportunities and uncertainties must be considered by these regulators:

- Changes in the way pipelines bid for gas and changes in their capital investment decisions
- Changes in the way costs are allocated and prices set for different classes of customers and for peak and off-peak use
- Increases in competition among pipelines for the unregulated industrial market and resulting changes in prices for jurisdictional sales
- Changes in the reliability of service provided by pipelines and distributors

Pipelines and producers face immediate questions of how and whether to renegotiate contracts and, together with the rest of the industry, they face longer term questions of how to arrange transactions in the presence of cost and demand uncertainty. Legislators and policymakers also have a legitimate interest in transactional arrangements, because of the close relationship between those arrangements and the efficiency of gas markets. But they also face broader questions of whether and how government should intervene to alter contractual relations agreed upon voluntarily by private parties.

Market rigidities introduced by contracts or other factors can have socially undesirable consequences, and natural gas is not the only market in which such rigidities appear. As Okun (1981) points out, such rigidities are pervasive and account in large part for the economy's poor handling of the food and energy supply shocks of the 1970s. Legislative bans or limitations on specific contract terms, such as current proposals to limit the operation of take or pay clauses, may fail to address deeper reasons for unresponsiveness of contractual obligations to transitory market shifts. Thus policymakers must reflect on whether it is desirable or possible to stimulate a more rapid movement toward flexible contract terms than private negotiations would bring about.

Fostering the development of new market institutions may also be a legitimate policy response to conditions in gas markets. The conflict between contract terms that provide incentives for efficient resource allocation and terms that provide for optimal risk sharing can be reduced only by greater stipulation and monitoring of each party's behavior under various contingencies. As field price deregulation brings greater uncertainty about prices and demand, contracts may thus become more complex—and costly to negotiate and monitor—and plagued with more problems. It is likely to be particularly difficult to design long-term contracts that transmit rapid changes in end-use demand to the upstream market to bring about equally rapid changes in supply. Overall, a move to other forms of transaction may be expected.

One way transactional arrangements may change is through the emergence of new market institutions within the industry. These may include the development of regional spot markets for pipeline and distribution company transactions—particularly off-system sales—and possibly a gas futures market as a way of hedging against price uncertainty. Contractual relations suffer, in comparison with an ideal competitive economy, because of the need to mute signals for changes in gas supply by terms that prevent all risks from falling on a single party. If it were possible to insure against price risks outside the contractual relationship, considerably better results could be obtained. One method of providing such insurance is the creation of an organized futures market in gas.

Another way to deal with contract problems in the face of price uncertainty is for downstream firms to integrate vertically upstream. Such developments could appear simultaneously with the growth of spot markets, creating a system of transactional arrangements more like that in unregulated intrastate markets. Vertical integration can reduce transaction costs and make the uncertainty associated with freer field markets easier to manage.

In intrastate markets, integration between producers and pipelines is more widespread than in interstate markets. Greater vertical integration in interstate markets would raise public policy issues about the competitive impacts of vertical integration. Indeed, most changes which might come about in downstream markets as a result of deregulation, and most policy initiatives dealing with those sectors, raise questions of what scope there is for competition in a natural gas industry with deregulated field prices.

Closely related to questions about the scope for competition are questions about the appropriate role for pipelines in the new gas market. Reformed or reduced regulation of pipelines, for example, could make it easier for gas to move to the regions and customers most willing to pay. But a move toward pipeline deregulation requires knowledge of the degree of interpipeline competition that exists or can be created. At the other extreme are proposals that pipelines should get out of the business of buying and selling gas, and act solely as transportation companies. Such a change could be motivated either by concern about pipelines' market power or by concern about pipeline motivation and behavior in negotiating contracts and choosing gas suppliers. Changing pipelines into price transportation companies would require new regulatory practices and, if common carrier status were to be imposed, probably new legislation.

These questions and policy issues have no immediate answers. Nor, save for some cautions about making policy without understanding the consequences, will answers be found in this study. But the issues are important enough to justify planning a systematic research program to obtain answers. It is to the methods and strategy of such research that we now turn.

3

Structural Changes in Downstream Markets Under Field Price Deregulation

Introduction

Issues pertaining to the competitive structure of the natural gas industry lie behind many of the problems of transactional arrangements and regulatory policy in downstream markets. Moreover, deregulation of natural gas field prices could bring about changes in the structure and state of competition in downstream markets that would in themselves warrant attention. Two possible changes may be in the degree of vertical integration (between pipelines and gas production) and in the degree of (horizontal) concentration in pipeline receiving and delivery markets. If vertical and/or horizontal restructuring of the transmission sector were to occur, it would directly affect the nature of interpipeline competition and thus the efficiency with which changes in supply and prices at the field level match with changes in demand and prices at the burner tip. To the extent that competition among pipelines is influenced by upstream deregulation, reform of the current regulatory and carrier status of pipelines as well as more direct action aimed at market structure may be necessary. (Analysis of the regulatory and carrier status of pipelines is taken up in chapter 6.)

Analyzing and evaluating the consequences of downstream regulation also requires an understanding of the state of competition in gas markets. All of the methods of analyzing the behavior of the regulated firm discussed in chapter 5 presume knowledge of demand elasticities facing that firm; these in turn depend on the degree of competition in that firm's market. An evaluation of changes in the regulatory status of pipelines, ranging from deregulation to imposition of common carrier status, also requires ways of measuring the degree of competition among pipelines.

This chapter provides an overview of the state of competition in gas production and pipeline markets and its implications for the economic performance, that is, allocative efficiency, of the gas industry. To develop methods that could be used in research about the performance of downstream gas markets and to understand what changes may come about in their structure, we apply the literature on industrial organization to two topics:

What structural changes can be expected to occur in response to the type of changes in the economic environment that result from field price deregulation

How the degree of interpipeline competition can be measured and studied

This brief survey of the literature on the relationship between market structure and economic performance provides some preliminary insights into potential policy issues that involve gas market structure and suggests topics that merit study.

Effects of Vertical Structural Changes on Downstream Competition

Relaxation of field price controls will change the configuration of risks and incentives for the industry's participants. Economic theory suggests that when downstream firms face uncertainty in input costs, they are likely to try to exercise greater control over upstream production (see Machlup and Tauber, 1960; and Hay, 1973). One possible way to gain greater vertical control is by fashioning different contractual arrangements; for instance, by devising provisions that shift risk away from downstream purchasers of inputs to upstream producers. However, the costs associated with a new contractual structure may outweigh the benefits, either because of the recontracting process itself or because costs of maintaining or governing the contractual relationship are high. Vertical integration—the joining of potentially separate successive stages of production under common ownership—may entail larger setup costs, but the greater control that ownership usually affords could provide offsetting benefits in the long run.

Vertical control by contract has been the predominant mode of transaction in natural gas, and it is unlikely that field price deregulation will change that. But because upstream deregulation may increase the cost of transactions and thus enhance the prospects for vertical integration, the mix of transactional arrangements could shift. In the intrastate gas market, which prior to NGPA was free of regulation at the federal level, there has been a more even mix of transactional arrangements. Vertical integration between pipelining and production, but also spot transactions, is more prevalent in the intrastate market compared with the interstate market. In the interstate market in the past few years, there has been a trend toward greater backward vertical integration by pipelines into production, and to a lesser extent by distribution companies (Cowan and Hagar, 1982, see chapter 2); spot transactions in the interstate market have been virtually nonexistent. Nonetheless, contracting is the norm in both interstate and intrastate markets, though as the intrastate market illustrates, the choice between contracting, integration, and spot transactions is not a mutually exclusive onc.

Of course, the extent to which vertical integration increases will depend on a number of factors besides transactions costs. The existing horizontal structure of pipeline markets is important.[1] A pipeline with monopsony power in a field market will derive fewer benefits from integration than one competing with many buyers (or fewer sellers). Institutional and regulatory rules may also constrain transactional behavior, and possibly "make or buy" choices—for example, whether to produce an input or purchase it on the market. Thus, for example, common carrier status for pipelines is likely to preclude *bona fide* vertical integration: a common carrier pipeline would no longer be taking title to the gas it transports. A more complete analysis of the determinants of, or the choices between, various transactional arrangements is contained in chapter 4. The changes predicted by that analysis should be evaluated in terms of the possible effects of increased vertical integration on competition in downstream gas markets.

At one level, the literature suggests that vertical integration will have a neutral impact on competition unless the vertically integrated firm attains a considerable degree of horizontal dominance in any one of the individual markets in which it operates, that is, the firm obtains a large share of market sales or purchases (see Bork, 1954). More generally, it is argued that integration will reduce economic welfare if the horizon-

1. We discuss later how upstream deregulation may change horizontal market structure and thus indirectly affect vertical structure.

tal structure of any of the vertically related markets is relatively concentrated, that is, relatively few firms account for a large share of sales (or purchases). An exception, in which integration can increase welfare, is if horizontal market power exists on *both* sides of some market (Schmalensee, 1973; Greenhut and Ohta, 1976).

The importance of the degree of market concentration is derived from the central theory in the study of industrial organization: The economic performance of a market is largely determined by the number and size distribution of firms operating in that market (the market's structural condition) and the pricing and other forms of conduct of those firms (see Scherer, 1980). When markets are relatively concentrated, firms are more likely to recognize their mutual dependence and be able to act to raise the market price above the competitive level. Depending on the industry's technological structure as well as other parameters, these firms can thwart potential competitors through "limit pricing" (Bain, 1956) or by making entry-deterring investments (Spence, 1977). Weiss's (1974) review of a large number of studies of the structure-conduct-performance relationship provides empirical evidence supporting the theory.

This line of reasoning suggests looking at available evidence on concentration in natural gas markets to get some idea of what impact an increase in vertical integration might have. Surprisingly, relatively little empirical research has been done applying this analytical framework to the natural gas industry in recent years. In the gas production segment there are two pieces of evidence, a paper by Schwartz and Wilson (1974) and a Federal Trade Commission (FTC) study by Mulholland (1979). Mulholland's analysis indicates that seller concentration levels in the production sector—based on either output or reserves—are "relatively moderate" and "similar to the median levels for the manufacturing sector and are below those threshold levels most commonly identified with monopolistic behavior." Schwartz and Wilson, on the other hand, argue that focusing solely on level of seller concentration masks other, perhaps more important structural indicators of the state of competition in gas production. In particular, they contend there may be significant market power realized by gas producers because the submarkets that comprise that portion of the industry are dominated by the largest petroleum firms, which engage in a web of joint ventures that may possibly be anticompetitive. Gas field markets, suggest Schwartz and Wilson, therefore may behave more like oligopolies.

In terms of the horizontal structure of the pipeline segment, there is to the best of our knowledge only one study using recent data. In a draft report on interpipeline competition for the Federal Energy Regulatory Commission (FERC), Mead (1981) suggests that while buyer concentration should not be a cause for concern, seller concentration may already be high. He concludes, after analyzing a number of pipeline markets, that there is a "moderate" amount of interpipeline competition in receiving (upstream) markets, but there is "very little competition" in the delivery (downstream) markets.

On the basis of just these three studies, it is difficult to make a definitive judgment on whether there are horizontal problems in gas markets, and hence, based on the "horizontal dominance" line of reasoning, on whether increases in pipeline vertical integration might be damaging to competition. Whereas at the production level it seems reasonable to place the burden of proof on those who seriously doubt some degree of "workable competition," there may be reason to be less sanguine about some pipeline delivery markets.

But the assessment of competition in a vertically organized industry such as natural gas must go beyond analysis of the horizontal structure of individual markets. Concentration ratios cannot tell the whole story. Meaningful judgments about competition require careful thought about market boundaries—in terms of geography and product definition, that is, degree of cross-elasticity of demand with potentially competitive products. The appropriate geographic boundary in gas production is an unresolved issue. Schwartz and Wilson argue that the field is the relevant geographic market, whereas Mulholland presents concentration calculations at both national and regional levels. The geographic boundary problem also exists in analysis of pipeline markets. Perhaps more important in pipelines than in production, however, is the question of product boundaries. Competition between natural gas and fuel oil may diminish the opportunity to exercise market power in otherwise concentrated pipeline markets, by making demand sufficiently elastic at the margin to frustrate attempts to raise prices by restricting output.

Boundary questions aside, market performance will also be affected by the height of barriers to entry that potential competitors face. Entry barriers are likely to be generally low in gas production. However, they are likely to be higher in newer producing areas, such as the Overthrust Belt or in offshore areas, to the extent that large initial capital outlays are necessary. Barriers to entry into transmission are, in part, created by regulatory franchise restrictions, but there are also economic factors, such as capital requirements for new construction.

The nature of regulation of local distribution companies also will influence the ability of pipelines to exercise any structural market power they might possess. Similarly, contractual agreements between pipelines and distributors will affect the scope of competition.

Finally, analysis of competition in an industry comprised of vertically related submarkets could be misleading if the focus is only on each individual market without any attempt to take into account its links with the markets upstream and/or downstream from it. When firms have multiple contact points across a number of vertically linked markets within the same industry, they may be more likely to recognize any interdependence of their pricing, output, or investment decisions—for example, when a small set of firms meet each other in the same input markets, in the same transportation markets, and then again in the same output markets. Broadman (1981) develops an analytical framework based on this notion and applies it to the petroleum industry.

Overall, the competitive effects of increased pipeline vertical integration arising from upstream deregulation could well depend on more subtle features of the industrial landscape than are revealed in an analysis of monopsony and monopoly power in receiving and delivery markets, respectively. On the other hand, integrated pipelines remain subject to FERC regulation. Blatantly anticompetitive behavior stemming from vertical integration—such as paying inflated transfer prices to affiliated gas producers and attempting to pass them on downstream—will be subject to institutional constraints on fraud and abuse. Questions of how pipeline regulation might be changed depend, at least in part, on the need for a regulatory system to impose such constraints.

Effects on Interpipeline Competition of Horizontal Structural Changes

Under NGPA there will be an uneven distribution of regulated, low-priced gas among pipelines. As a result, there will be major differences in the average cost of gas among pipelines simply because they have different proportions of "old" (cheap) and "new" (expensive) gas. In general, intrastate carriers will have higher average input costs because most of their gas will be priced at the margin—that is, almost all of their gas will not be subject to regulation. The position of interstate pipelines, however, will be mixed. Some interstate companies will most likely obtain up to 80 percent of their throughput at controlled prices, but other interstate pipelines could be carrying as much as 70 percent decontrolled gas (see EIA, 1981).

The short-run consequences of these differentials in cheap gas "cushions" are relatively straightforward. A pipeline with a small cushion will be forced to pass on to its customers higher average gas costs. Two structural adjustments could result. First, to the extent that burner tip gas prices increase relative to alternative fuels, the pipeline will lose customers. In the long run, as the pipeline's load is diminished, unless it reduces its capital stock, its performance, that is, its rate of return, will begin to decline. [This assumes the pipeline utilizes purchase gas adjustment clauses (PGAs) to pass on increased costs.]

The second adjustment may result from competition by larger cushioned pipelines which can offer lower cost gas. This will also reduce the market share of the original pipeline. Shifts in market shares can take place in the short run even though pipeline certificates of convenience are tantamount to bestowing a quasi-franchise because a number of markets are served by several trunk lines. In the longer run, in a situation where there is a deep cushion differential, it may even be economic for a large-cushioned pipeline that otherwise would not be a potential competitor to expand into new geographic markets.

These two adjustments have differing impacts on consumers in the market in question. Rising gas prices make customers of low cushion pipelines worse off, and these customers limit their losses by switching consumption to an alternative fuel. With the adjustment brought about by interpipeline competition, however, customers may benefit from costs that are lower than if they remained on the small-cushioned system. If pipeline conduct is unaffected by changes in market share, interpipeline consumption and switching of customers could eventually result in uniform nationwide prices (save for true delivery cost differences), after a possibly painful transition. But even this benefit may be unachievable. To the extent that regulation is imperfect, firms with larger market shares have better opportunities to exploit the potential market power that results from a greater degree of horizontal dominance (see Shepherd, 1972).

Up to this point the discussion has focused on market structural changes induced by NGPA that will be regional and/or pipeline-specific. But holding these characteristics of the potential changes constant, there are also likely to be structural changes across customer classes. As a practice, downstream gas firms have an incentive to discriminate their rate structures in accordance with demand elasticities. (Analysis of pricing behavior is the subject of chapter 5.) With large gas cushion differentials, pipelines with small cushions will attempt to concentrate as much of their costs as possible on customers with low demand elasticities, subject, of course, to the institutional constraints governing cost allocation among customer classes (see chapter 6). Small-cushioned pipelines, other things being equal, will tend toward supplying inelastic markets and large cushioned pipelines, toward more elastic markets.

An equilibrium will come about where, in seeking additional supplies to meet their increased sales, large-cushioned pipelines will purchase increasingly costly gas, and small-cushioned pipelines, though they ultimately will lose sales, will attempt to reduce input costs by buying a smaller proportion of unregulated gas. The latter objective could be achieved by (1) trying to obtain cheaper gas through interpipeline sales or through acquisition of production entities with low-cost reserves (that is, vertical integration), and (2) not honoring contracts for high-cost gas.

In the long run, of course, the cushion differential will disappear as the supply of regulated gas is reduced. Yet in the process there are likely to be major performance differences among pipelines with different sized cushions and possibly severe financial hardships. The result will be a change in the horizontal structure of various pipeline markets. As noted above, industrial organization analysis suggests this could affect the degree of competition in these markets and ultimately the level of gas prices at the burner tip.

Two additional pieces of information should be noted in this context. First, there are already likely to be incentives endemic to utility-type regulation for greater horizontal integration than otherwise would occur. Peles and Sheshinski (1976) show that it is advantageous for two or more otherwise separate rate-of-return regulated utilities to merge horizontally, say across geographic markets. Such "market-extension" integration affords the regulated firm greater flexibility in making investments that increase the level of permitted profits. Second, as Mead has argued, some pipeline markets may already lack effective competition. Large cushion differentials could exacerbate concentration in these markets.

Concluding Observations

Economic theory provides reasonably strong clues as to how resources will be misallocated in the presence of market power. But analysis of the impact of upstream deregulation of natural gas on the competitive structure of downstream markets is complicated by the vertical organization of the industry and the effects of regulatory and institutional constraints that will continue to affect the transmission and distribution sectors. The results of standard models of industrial organization need to be modified to take these factors into account. Existing studies of downstream market structure in gas can provide only a rough assessment of how interpipeline competition might be affected by the relaxation of field price controls. This conclusion is important in considering proposals to reform pipeline regulation based on presumptions of the degree of pipeline monopsony power in field markets and/or monopoly power downstream (see chapter 6).

The NGPA can be expected to cause some horizontal structural changes in downstream markets. Under NGPA, regulated (low-priced) gas will be unevenly distributed among transmission companies. As a result, in the short run there will be major gas cost and hence major performance differences among pipelines simply because they have different proportions of "old" and "new" gas supplies. In the long run, adjustments to the uneven distribution of cheap gas could bring about a significant change in market shares within the pipeline industry and possibly a change in the degree of competition between pipelines. This, in turn, could affect prices at the burner tip.

Whether changes in market share pose problems depends in part on the current state of competition in gas markets, as does the evaluation of proposals for more or less stringent pipeline regulation. In field markets, the available evidence indicates workably competitive conditions, while analysis of the pipeline sector suggests that some markets—notably delivery (downstream) markets—may not be sufficiently competitive. In both cases, however, serious problems arise in the definition of product and geographic market boundaries, and the literature fails to incorporate measures of competition other than simple concentration ratios. As a result, these conclusions are only tentative. Further research defining relevant market boundaries in the gas industry, analyzing the effects of regulation, and using more sophisticated measures of competition is thus warranted.

4

The Impact of Field Price Deregulation on Vertical Transactional Arrangements

Introduction

Vertically related activities can be coordinated in a number of ways, ranging from market exchanges between anonymous parties who never meet again to centralized direction of permanently related divisions of an integrated firm. Contractual relations occupy an intermediate position, as they involve independent parties but also specify how each shall behave and be rewarded during some ensuing period. These different ways of conducting transactions can coexist within an industry. Diverse arrangements make it possible for traders to choose ways of buying and selling gas best suited to these particular circumstances. But the balance among transactional arrangements may be changed by external events, such as field price deregulation.

Production and long-distance transportation of natural gas began as highly integrated activities, but passage of the Natural Gas Act (NGA) in 1938 (which subjected interstate transmission of gas to federal regulation) and expectations that the act would not extend to upstream activity gave rise to gas production as a separate industry. With the proliferation of "independent" gas producers, an increasing percentage of transactions between producers and pipelines were made under long-term contracts. Spot markets, involving immediate transfer of a single packet of gas, could not develop, whatever their technical possibility, because of procedures for pipeline certification and financing that required long-run assurance of input supplies.

The 1954 *Phillips* decision brought production of gas sold to interstate pipelines under federal regulation and solidified the existing mixture of vertical transactional arrangements. That mixture is still prevalent today in an industry comprised of, on one hand, unintegrated producers, transmission companies, and distribution companies all dealing exclusively through contractual arrangements, and on the other hand, various vertically integrated combinations of these three segments. All integrated companies also rely on contractual arrangements to varying degrees. The predominant mode of transaction remains the long-term contract, although a significant trend toward integration has become apparent, and there are an increasing number of "spot" transactions in the form of off-system sales.

Deregulation of field prices is likely to require changes in the way in which transactions between producers, pipelines, and distributors are arranged. The long-term contracts which largely govern transactions within the industry contain terms and provisions that seem ill-suited to current market conditions. Contracts that specify in advance the price and quantity of gas to be moved can prevent efficient adjustment of supplies to meet unforeseen movements in demand. Contracting practices held over from an era of stable prices and increasing demand may also allocate the risks created by cost and demand uncertainty in an arbitrary and inefficient way.

Problems of disequilibrium and excess cost in the contracting process led in 1982 to a number of proposals for legislative action to alter or constrain contract terms. These ranged from limiting take or pay provisions to allowing gas purchasers a single opportunity to renegotiate contracts. It is necessary to analyze how and why contract terms will change through voluntary renegotiation, and how well the result comports with public interest in order to evaluate the need for intervention in the contracting process.

Even if government does not intervene in the contracting process, actors in the industry seem unable to agree on what changes in transactional arrangements would be in their mutual interest. The related questions of how (and whether) workable gas supply contracts can be written and how the choice between contracting and alternative transactional arrangements should be made have not been addressed systematically in studies of natural gas markets. To develop analytical methods that can provide answers to these questions, it is necessary to delve deeply into new theoretical developments in the field of industrial organization. Some particularly useful approaches to the choice of transactional arrangements are found in the growing literature on the economics of information and uncertainty, particularly that which applies to the comparison between market and nonmarket means of resource allocation. These broad questions of how to organize transactions are addressed first in this chapter.

For two reasons, detailed examination of long-term contracting seems justified. First, the possibility of improving the contracting process must weigh heavily in the choice between continued use of long-term contracts and some alternative approach. Second, it is unlikely that the use of long-term contracts will be abandoned in the foreseeable future. Thus the second part of this chapter develops methods for analyzing the structure of natural gas contracts, based on developments in the literature on principal-agent relations, and on optimal incentive contracting. Because the fields of economics referred to in this chapter are both new and somewhat abstruse, we devote much of this chapter to developing some basic concepts.

Examples are then given of how these concepts can be applied to analysis of natural gas markets. The broad choice among integration, long-term contracting, and spot transactions is approached with an analytical framework predicated on the notion that the form of vertical transactional arrangement is largely determined by transaction costs. This literature suggests that when uncertainty about supply and demand is combined with the fixed costs characteristic of gas production and transmission, conditions unfavorable to market transactions are created.

Thus field price deregulation, by increasing uncertainty in gas markets, could stimulate a movement toward greater vertical integration. However, the advantages of integration in gas markets are not absolute, and contracting processes can evolve to a form more suited to an uncertain environment. An assessment of how the proportion of contractual to integrated activities will change requires more detailed analysis of the functions and defects of long-term natural gas contracts. Such analysis can be based on methods developed in studies of "principal-agent" problems and related investigations of optimal incentive contracts. We use this literature to suggest methods for predicting and evaluating likely changes in the nature of natural gas contracts, which are likely to continue to coexist with vertically integrated activities.

Two additional possibilities are considered in this examination of potential changes in transactional arrangements. One is that institutional and regulatory factors may play a significant role in determining how transactions are carried out.

They may do so directly, by constraining choices that firms may make, and indirectly by altering risks and transaction costs.

Another possibility is that entirely new market institutions will develop, thus leading to a menu of possible transactional arrangements more complex than the simple progression from spot to integrated exchange. Introduction of future markets or similar devices to reallocate risks and increase information is an example of such a development.

Relative Costs of Transactional Arrangements

Systematic discussion of the determinants of vertical transactional arrangements starts with Coase's (1952) analysis of the nature of the firm. Coase recognizes that firms will bypass the market in order to obtain inputs or to distribute output if the costs involved in carrying out purchases or sales between two or more parties in open markets are higher than can be achieved internally. Coase's argument applies both to special conditions of technological coordination where the incentive for nonmarket transactional arrangements is perhaps obvious, and more generally. An example of the former is the savings in fuel costs made possible by the direct transfer of molten pig iron from the blast furnace into the steel converter.

Williamson (1979) deepens Coase's analysis of transaction costs and the boundaries of the firm by going beyond the simple "make or buy" decision. His major thrust is that the cost of using the market is likely to be substantial if there is "inherent uncertainty" about future market conditions or if there are a small number of firms in the market. Under these conditions, according to Williamson, spot market transactions—unconditional purchases or sales made immediately and often settled with cash—will be less attractive than arrangements which either entail contractual obligations or ownership of input production or output distribution facilities. But there are significant differences in transaction costs between contracting and integrating, and it is here that Williamson's analysis is most applicable. He distinguishes between transactions that are occasional and those that are recurrent and between transactions involving relatively standardized assets and those involving specialized assets. In addition, three components of the cost of contracting are identified: the costs incurred in arranging contractual relationships (the search costs of finding contractual partners and negotiation costs), the costs of contract monitoring, and the costs due to "opportunistic" behavior by another party (or parties). These costs are likely to be low when transactions involve relatively standard assets and occur occasionally or at irregular intervals. In these circumstances firms will tend toward contracts that specify in advance what will be done and what will be paid. But with recurrent transactions involving highly specialized physical (or human) assets—specialized in terms of technical characteristics, location, and so on—these costs are likely to be higher. Centralized internal management is likely to be more cost-effective than market exchange under those conditions.

Klein, Crawford, and Alchian (1978) focus more closely on postcontractual opportunistic behavior as a constraint on interfirm trading. They emphasize that once a specific investment is made, quasi-rents are created and the payoff from reneging on contracts is likely to increase. As threats to the security of contracting become more credible—as they will, once a given asset becomes more and more specialized—the cost of contracting increases relative to the cost of intrafirm trading. Klein, Crawford, and Alchian argue that in their attempt to reduce the costs of avoiding the risk of appropriation of these quasi-rents (by "opportunistic" parties), otherwise contractually related firms will find it mutually advantageous to own jointly the specialized asset(s).

While Williamson and Klein et al. are certainly correct in pinpointing postcontractual opportunistic behavior as an important component of the transaction costs of contracting, relatively less attention has been paid to the importance of goodwill as an enforcement mechanism of the contractual process. Firms generally strive for long-term business relationships to avoid incurring prospective transaction costs of finding new business and negotiating contracts that can subsequently be exploited or broken. On the other hand, contracts may be enforced more formally by making payment contingent on performance. Klein and Leffler (1981), for example, argue that potential contract cheaters could be offered a stream of payments over the life of the contract that is at least equal to the potential increase in wealth gained from cheating. The equilibrium level of such contract premiums, or more generally the nature of a given contract's provisions, will be determined by how symmetrically information is distributed among the involved parties and by their attitudes toward risk. This analysis shades off into the literature on optimal contracts, which recognizes that perfect incentives may be unattainable and then searches for a contract design that supports the best achievable outcome. This approach to the issue, viewing contractual parties in the context of a principal-agent relationship, is pursued below. Finally, beyond the question of goodwill or the use of side payments, opportunistic behavior will be constrained if markets are competitive. In this sense, the relative cost of contracting is a function of market structure.

An equally important factor influencing the relative cost of contracting in the presence of uncertainty is the more fundamental element of the limits of human rationality. Drawing from Radner (1970), Williamson (1973, 1979) argues that comprehensive contracting—comprehensive in the sense of attaining the level of welfare consonant with that of a competitive equilibrium—is unobtainable because of the limited capacity of individuals to "receive, store, retrieve, and process information without error" (Williamson, 1973). In essence, the point is that it is virtually impossible to put into contracts provisions adequate to guarantee a competitive solution. The relevant issue then becomes the extent to which integration in the presence of uncertainty can improve welfare.

Another cost-reducing alternative to a long-term contract is a sequence of short-term contracts that are renegotiated at frequent intervals. By contracting for a shorter time, parties can reduce the number of contingencies for which rights and duties must be specified. A renegotiation makes it possible to update contracts on the basis of market developments rather than specifying all conceivable market developments in advance. However, recontracting may also introduce the chance for "opportunistic" behavior. Crawford (1982) finds that if one party to a contract must invest in "relationship specific" capital, short-term contracts make it possible for the other party to exploit the increased vulnerability created by that investment. Thus the party who must invest in relationship-specific capital will not fare as well under a sequence of short-term contracts as under long-term contracts, and will be biased against making such investments. Under these conditions a sequence of short-term contracts can fail to achieve as high a level of welfare as a single long-term contract.

Contracting versus Integration: Theoretical Rationales

As mentioned in the previous chapter, firms may integrate in order to avoid the inefficiencies associated with monopoly or monopsony in the field market (Machlup and Tauber, 1960; Hay, 1973). Deregulation could thus encourage integration if either pipelines or producers possess significant market power, and such integration could reduce the efficiency of gas markets.[1] But it is not necessary to assume that field markets are imperfectly competitive in order to conclude that deregulation will create a trend to greater integration. The incentives for integration would come from problems of information and

1. The incentive for vertical integration and the consequences of integration for consumers also depend on the elasticity of input substitution. Major work in this area includes Burstein (1960); Vernon and Graham (1971); Schmalensee (1973); and Greenhut and Ohta (1976). Briefly, if vertical integration leaves the cost conditions associated with each stage in the production process unchanged, and inputs priced above marginal cost are combined in fixed proportions to produce a unit of final output sold in a competitive market, that is, an upstream monopolist with a Leontieff production technology integrates into a competitive downstream market, the price of the final product will not be affected. Under the assumption of variable input proportions, vertical integration under these market conditions will result in a lower price in the final market. However, if the integrated input monopolist gains control over the downstream market's use of all the (substitutable) inputs, it is likely to raise final product prices (see Warren-Boulton, 1978).

coordination in market-mediated exchanges. In this case integration might serve a beneficial economic function.

Several writers have hypothesized that vertical integration is a response to uncertainty. Most prominent among these works are (1) Wu's (1964) analysis of integration as a way of avoiding inefficient response to fluctuating input demands; (2) Green's (1974) study demonstrating that price rigidities in intermediate product markets can be overcome by integration; and (3) Arrow's (1975) argument that vertical integration will lessen the impacts of fluctuations in input supply prices when information between participants at the upstream and downstream stages of production is unequally distributed.

Arrow's model is most insightful. The supply of each upstream firm is taken to be a random variable and is not subject to managerial control. Each upstream producer knows one period in advance what its supply will be, but downstream firms know only the probability distribution of supplies. While there is a spot market for the upstream product (one of the downstream firms' inputs), capital (the other downstream input) must be chosen one period in advance. Downstream operators are shown to have an incentive to engage in upstream production (through acquiring upstream firms) because this will improve their ability to predict the (spot) price of the upstream product and thus their ability to select the level of capital. In fact, Arrow's model leads to the conclusion that it will be advantageous, under these conditions, for each downstream firm to acquire more and more upstream firms, so that in the limit the entire input market will be owned by one vertically integrated firm. Hence, although initial conditions were competitive, imperfect competition is an inevitable outcome in Arrow's model. Arrow emphasizes that the purpose of integration is to acquire information, not to gain market power or a more certain supply of inputs.

Blair and Kaserman (1978) show that in the presence of uncertainty an input monopolist selling to a competitive downstream market characterized by a fixed-proportions production function has an incentive to integrate forward as long as risk preferences are not linear or identical between stages. They argue that vertical integration, by reallocating risk to those firms most willing to accept it, acts as a form of insurance but without the moral hazards. (The concept of moral hazard is described below.)

Williamson (1979) has also made a similar point. Under a contractual arrangement, should the upstream supplier be willing to bear the risks of producing a specified product of uncertain cost, a fixed-price contract containing a risk premium to the output price could be offered. However, the downstream buyer might find this premium too high and be willing to take on the risks and offer a cost-plus contract. Of course, there is the problem of monitoring the supplier's performance to ensure that costs are minimized. Vertical integration is thus viewed as a way to overcome the moral hazards affecting supplier behavior and thus would make such a contract attractive. As Williamson notes, internalization of these risks through vertical integration does not reduce the need for monitoring the production process, but such monitoring presumably can be done more cost effectively with internal resources rather than through an external agent.

Carlton (1979) also presents a model of the transmission of uncertainty between a product market and one of its factor markets. Carlton's model exposes the incentives for and consequences of (backward) vertical integration. The key feature of the model is that uncertainty about being able to sell all of their (unstorable) production makes it necessary for upstream suppliers to charge a price higher than marginal cost. A downstream purchaser who knows that some quantity of inputs will be needed with high probability can acquire them at lower cost by making than by purchasing. Remaining input demand will be purchased if needed from independent suppliers—who will find themselves facing even more uncertain demand and thus be forced to charge even higher prices to recoup losses on unsold products. Thus backward integration can be expected because it reduces average costs of the downstream firm, even though it increases the risks faced and prices charged by independent upstream firms.

There would be a similar result if the downstream firm could offer to pay a low price (equal to cost of production) for firm sales and then a higher price for spot purchases to meet high demand. Carlton shows conditions under which the increase in the integrated firm's profit that results from *making* the first unit of input exceeds the increased cost of all additional *purchased* units and thus would make such a contract attractive. As there is nothing in Carlton's model that prevents a contractual relationship with variable prices, Carlton's analysis does not clearly discriminate between integration and contracting. It does suggest there are reasons to expect integration rather than spot sales under the conditions that characterize gas markets.

The transactions cost theory of nonmarket vertical arrangements is helpful in explaining why contractual relations are chosen in some circumstances and why integration is chosen in others. But the primary drawback of the theory is that the notion of transactions costs is not defined with sufficient clarity to allow these costs to be measured. Consequently some care is required to avoid circular reasoning in which high transaction costs are inferred from the existence of some market structure and then used in turn to explain that structure. Indeed this explains why the theory has not been directly tested empirically.

Moreover, as Flaherty (1981) has pointed out, the theory is not robust enough to distinguish vertical control—which can come about through long-term contracts as well as through integration—from vertical financial integration. Flaherty presents a theory of financial integration and a consistent, but separate, theory of the control of transactions in a vertical product flow. Drawing on Weitzman's (1974) theory of the differences between price and quantity controls in central planning, Flaherty shows that in many industrial markets quantity instruments are more efficient than price mechanisms in coordinating upstream activity with changes in demand when downstream operators have incomplete information about upstream costs, and this applies whether or not the upstream and downstream firms are financially integrated. By superimposing Williamson's explanation of transactional arrangements on the prices versus quantities theory, Flaherty is able to predict for many special cases when firms that use quantity controls will integrate and when they will remain financially separate. Briefly, assume a situation where there is likely to be a long-run relationship between downstream and upstream segments, and downstream production uses inputs in fixed proportion. If it is expected that the relationship between upstream and downstream units will require lengthy negotiation because the downstream unit is unable to learn about upstream production costs (and thus is incapable of knowing whether it is getting cheated), then financial vertical integration could solve the market failure by giving both units claim to joint profits.

Implications

The theoretical treatments of spot transactions, contracting, and integration discussed above suggest that field price deregulation could create conditions favorable to increased vertical integration. The greater uncertainty that deregulation brings to price and demand combines with technical features of gas production and distribution to bring about these conditions. Arrow (1975) finds that differences in information, similar to those likely to exist between pipelines and producers, also create an incentive for integration to improve coordination of vertically related activities. This result is more suggestive than definitive, because Arrow analyzes a case of uncertainty about upstream supply whereas in natural gas uncertainty about downstream demand is at least equally important.

Carlton's analysis is pertinent to the gas market, with some reinterpretation. The gas producer's fixed investment in drilling and equipping must be amortized in the selling price of gas, so that uncertainty about marketability raises the break-even price. (This also helps explain take or pay provisions, which can allow a lower price per unit by guaranteeing purchase of more units.) A gas pipeline can lower its gas acquisition costs by producing some gas for its own account to supply high probability demand or by agreeing to take or pay contracts.

Conditions favorable to opportunistic behavior, stressed by Williamson and by Klein, Crawford and Alchian as a reason for integration, may also exist. With few exceptions, natural gas can be transported from the wellhead to trunklines only if a fixed investment in gathering lines is made. Once made, this investment binds the two parties to a transaction together, and gives one or both parties a quasi-monopoly status.

Problems of sharing risk and monitoring compliance with contracts are seen by Williamson (1979), Blair and Kaserman (1978), and Flaherty (1981) to create a preference for integration under these conditions of uncertainty and mutual dependence. However, transaction costs associated with contracting must be compared with the cost of providing incentives and monitoring performance within a vertically integrated firm (see Spence, 1975a).

The cited studies suggest the general conclusion that introducing greater uncertainty into a contractual relationship requires an increasingly complex agreement that will be more costly to negotiate and monitor. The simpler a contract is, the less well it will motivate parties to adapt their actions to exogenous events. When those costs of contracting are considered, increasing uncertainty thus reduces the efficiency of contractual resource allocation relative to the efficiency of resource allocation within a firm. If costs of internal control increase with the scale of integration, a balance will be struck between buying gas and producing internally. Thus vertical integration may increase with field price deregulation, but long-term contracts will not completely disappear from use. How far the substitution of integration for contracting will proceed depends in part on what can be done to adapt the contracting process to conditions of greater uncertainty. Answering this question requires more detailed analysis of contractual incentives and costs.

Moreover, since long-term contracts are likely to remain in use, the future efficiency of gas markets depends critically on improvements in the performance of gas contracts. Understanding the extent to which the contracting process can be improved, predicting likely directions of change, and evaluating interventions in the contracting process also requires more detailed analysis of the contractual relationship.

Arranging Transactions Through Contracts

It is not surprising that contracts formulated during an era of wellhead price controls might be unsuitable to a deregulated market. However, an examination of contractual relationships suggests that no easy solution exists. A contract determines the risks and rewards faced by each party and it creates incentives for parties to behave in particular ways. In order to assess these incentives, it is necessary to formulate an explicit model, in which each party's maximizing behavior is affected by contractual constraints. Such models have revealed, in general, a conflict between risk-sharing and incentives: provisions which support an efficient allocation of risk will also create incentives for inefficient use of real resources, and incentives for efficient resource use may also produce an inefficient allocation of risk.

The "theory of agency" provides the simplest examples of how such models can be developed and analyzed. Describing the abstract principal-agent problem makes it possible to identify the elements that must go into a general bargaining model of natural gas contracts.

Concepts

The "theory of agency" was developed independently by Robert Wilson (1968), Stephen Ross (1973), and Michael Spence and Richard Zeckhauser (1971). Wilson's work covers topics similar to those of Ross, discussed below. Spence and Zeckhauser develop a beautifully clear analysis of how limited information produces a conflict between the need to share risks and the need to provide incentives. Their analysis is based on the concept of "moral hazard," typified by a driver who, being fully insured, drives less safely than he would if he were personally liable for damages. Spence and Zeckhauser provide examples from medical insurance in which the insurer pays an amount determined by the insured's choice of treatment. If the insurer cannot observe the patient's condition directly, some coinsurance payment is required to prevent the patient from choosing the most expensive possible treatment. The higher the coinsurance rate, the more likely the patient is to equate the marginal cost of treatment with its marginal benefit. But high coinsurance rates defeat the whole purpose of insurance, which is the pooling of risks to reduce the uncertainty about medical bills facing any individual. The coinsurance rate (and premium or deductible provisions) must be set at a level which allows the insurer to break even on average. Subject to this constraint, the optimal coinsurance rate is one which balances the loss in expected utility that comes from greater uncertainty about medical costs against the gain that comes from avoiding excessively costly forms of treatment.

This tradeoff can be made less harsh if the insurer can obtain information about the condition of the insured. Ideally, if the patient's condition could be observed by the insurer, a system could be devised in which the insurer paid only for standard treatment identical to that which an uninsured patient would have chosen. Similar problems of reconciling motivation with risk-sharing, and similar results on the value of information, appear in the more complex content of natural gas contracts.

Ross (1973) launched general consideration of the "principal-agent" problem by raising the question of how differences in attitudes toward risk introduce inefficiencies into a relation between the principal (for example, the medical insurance company) and the agent (the patient). Ross considers a very simple relationship. The principal receives a net return that depends on the action taken by another party, the agent, and on some exogenous random variable. The problem can also be stated in terms of a principal who will receive the net proceeds of some investment. There is a range of possible investments, each with a different probability distribution of net returns. The agent chooses which investment the principal will make, and is compensated according to some agreed-upon fee arrangement.

Whatever fee is paid the agent, efficiency requires that the principal's most desired act be undertaken, which the agent will only do if its attitude toward risk is properly related to the principal. If, for example, the agent were much more risk averse than the principal, the agent might choose a safe investment in order to obtain a fee with low variance, while the principal could prefer a riskier choice. Thus the best outcome possible will be a compromise, providing lower expected utility to the principal than he could achieve by choosing an investment himself and paying the agent sufficient compensation to keep the agent's expected utility constant. Following this line of reasoning, differences in risk preference between consumers and suppliers of gas could lead to security and predictability of supply that is greater or less than optimal.

Ross does not treat the problem of providing an incentive for the agent to exert an adequate level of effort, when that effort is costly and cannot be observed directly, nor does he consider any fee arrangement other than sharing of proceeds. Shavell (1979) and Harris and Raviv (1978) investigate this problem in papers that apply the methods of principal agent theory to devising optimal incentive contracts. They establish three results of great importance to gas contracts and provide a method of analyzing features of these contracts. The results are:

> If both parties are risk averse, and information is differentially distributed, an optimal contract must share risk between the parties.
>
> Under these conditions, both parties would be better off under ideal central direction of their activities than they can be made by even the best-designed contract.
>
> Greater information on the actions of the agent makes it possible to write a more complex contract that can increase the welfare of both parties.

Shavell approaches modelling a principal-agent relationship by specifying the mechanism that generates risks and payoffs, the information sets and risk preferences of each party, and a definition of optimality. The principal's utility depends on some outcome, which has a probability distribution conditioned on the effort taken by the agent, and on the payments he makes to the agent. Compensation to the agent is assumed to depend either on the observed outcome alone or on the outcome and some uncertain indicator that is imperfectly correlated with the agent's effort. The agent takes as given the compensation he will receive as a function of outcomes and indicators of his effort (the fee schedule), and chooses an action that maximizes his expected utility subject to that fee schedule.

To find a Pareto optimal fee schedule, Shavell constrains the agent's utility to be no less than some prespecified value, and chooses the fee schedule that maximizes the principal's expected utility subject to that constraint. The fee schedule affects the principal's utility through its influence on the agent's effort, which determines the probability distribution of outcomes, and through the payments it requires.

A first-best optimum, achievable by integrated management, is found by choosing *both* the agent's compensation *and* the agent's effort to maximize the principal's utility, subject to a constraint on the agent's utility. Shavell shows that it is always possible to make both parties better off by integration except in two extreme cases. Only if the principal can observe the agent's action perfectly *or* if the agent is risk neutral will the agent's choice of action under the incentives provided by a Pareto-optimal fee schedule be that which achieves the first-best optimum.

If the outcome is determined uniquely by the agent's action, so that there is no uncertainty, the first-best optimum can also be achieved because observing the outcome provides the principal with perfect information about the agent's action. (This point is not made by Shavell, but is implicit in his results.) Otherwise the first best outcome can only be achieved by joint management and direction of each party's action; the outcome from a Pareto optimal fee schedule is never as good, but is the best achievable when motivation and compensation for risk bearing are provided by one fee schedule.

Shavell demonstrates why risk should be shared in a Pareto optimal contract. The need to spur effort requires that the agent's fee must depend to some extent on the outcome, but if the agent is risk averse, it must never bear all risk. As a result, the agent's effort will be less than optimal, compared with what it would be if the principal could observe and control (or reward) the agent's effort directly. Shavell demonstrates the need for risk sharing by noting that if the agent's compensation is independent of its action, it will have no motivation to exert itself. A small increase from zero in the agent's risk has no first-order effect on the agent's utility, but will increase the agent's effort and the principal's expected utility. If the agent bears all risk, returning some risk to the principal will not reduce the principal's expected utility but will increase the agent's (due to a reduction in risk-bearing). The small change in risk will have no first-order effect on the agent's effort, so that principal and agent can be made better off by shifting some risk to the principal and reducing the fee schedule.

Shavell also shows that information about the agent's effort is always valuable, making it possible to construct a fee schedule depending on outcomes and information about effort that improves the positions of both principal and agent, relative to what could be achieved with no such information. But only perfect information or the existence of a risk-neutral agent makes overall optimality possible. The value of information also depends on the efficiency of the agent's effort. If, Shavell argues, little effort is required to change the outcome drastically, "only a slight deviation of the fee schedule from a first-best schedule is needed to overcome a problem of an incorrect incentive to take effort" (Shavell, 1979).

How closely a contract will approximate a first-best optimum also depends on the cost of obtaining information about the agent's effort. The more costly information is, the less will be acquired and the greater will be the deviation from a first-best schedule. The value and use of information in designing optimal incentives is explored in more detail by Holstrom (1979). These results suggest that in the uncertain environment created by field price regulation, there will be movement toward more monitoring of what the parties do and more complex contracts using that information to condition obligations and rewards.

Harris and Raviv approach the problem of optimal contracting in a manner similar to Shavell. They show that what variables are included in a contract depend on what both parties to the contract are able to observe,[2] and that if all rel-

2. Operationally, only variables observable by both parties can be used to trigger actions or obligations in the contract.

evant information is available to both parties, then the choice of contract terms is independent of attitudes toward risk. In this case, a "forcing" contract is possible: for each state of the world, the principal determines what the agent's optimal effort would be, and requires the agent to expend at least that level of effort or receive no fee (or pay a large penalty). The fee can depend on the outcome in a way that shares risk optimally. Unfortunately, this propitious outcome would require an impossibly complex contract in order to state required actions under every possible contingency.

Shavell and Harris and Raviv both argue that if the agent is a risk-neutral agent and can fully insure the principal's risks, then a "first-best" contract can be written. The agent can offer the principal a fixed payoff, and bear all risk. In the pipeline/producer case, this might be a contract in which all demand or price uncertainty would be borne by the producer. Such a contract would place the pipeline in the role of a transportation company that is paid a fee for transport services by the producer selling gas directly to the pipeline's former customers. Such an arrangement cannot be optimal if the producer is risk averse, though it does avoid the moral hazard problems of motivating producer's effort. Ross (1976) also shows that if the principal has a choice of employing one of two agents whose effort can be observed, the one more willing to take risks will always be preferred. The way the proposition is proved also points up the cost of dealing with a risk-averse agent. Any service that the more risk-averse agent would provide for some fee will be provided by the less risk-averse agent for a smaller fee, for the less risk averse agent requires smaller compensation for bearing risk. Accordingly, the benefits of a contract to the principal become smaller as the agent becomes more risk averse.

Implications

The central insight of the principal-agent literature might be called the "coinsurance and deductible theorem": if the principal cannot observe what effort the agent is expending, the best achievable contract with a risk-averse agent is one in which the agent bears some but not all of the risks associated with the outcome of his actions. A contract that shares some but not all risk could be one in which the agent is paid a fixed fee (analogous to the deductible) plus a payment dependent on the outcome of concern to the principal (analogous to coinsurance). This "two-part pricing"—rather than a simple sharing of proceeds or single price for gas—will be seen in later chapters to offer many advantages in coordinating investment and output decisions.

Principal-agent theory also suggests conditions that make contracting an imperfect transactional arrangement. If it is impossible to observe all relevant external factors that affect outcomes, or to observe the agent's actions directly, it is also impossible to achieve the same expected utility that an integrated decision process would achieve. The less information is available or utilized, the lower is the expected utility achieved. When perfect information is not available, how well a contract will perform also depends on the attitudes of the two parties toward risk, on how costly it is to obtain information, and on how naturally their interests fall in line.

Applications to Analysis of Natural Gas Contracts

Natural gas contracts contain a bewildering array of terms and provisions, and complete analysis of the contractual process requires extensive examination of their nature and significance. To be manageable, that analysis should proceed incrementally by characterizing the operation of a few major provisions and then recognizing how their effect is altered by subsidiary provisions. This study can only illustrate an approach to that analysis.

During 1982 attention focused on two key provisions found in many gas contracts: price escalators that provided for increases in gas prices independent of supply and demand conditions in the market, and take or pay provisions that required pipelines to accept delivery of or pay for some minimum amount of gas. Similar terms appear in contracts between pipelines and distributors as well as in contracts between pipelines and producers. In some cases, these provisions are modified in significant ways, by buyer protection clauses that allow renegotiation of prices or delivery levels under some circumstances, by grace periods in which required deliveries can be made up, and by requirements that pipelines transport to others gas that they refuse to buy.

Price and delivery provisions are central to any long-term contract: neither risk allocation nor incentives can be determined unless *both* are defined. Those provisions shape the behavior of both parties and the success of any given contractual relationship. To analyze how contract terms are chosen, and how increased uncertainty may change contracts, it is necessary at minimum to examine alternative methods of specifying price and output levels, and to see whether they create incentives and an allocation of risk preferable to both parties. This extended analysis requires a description of the sources of uncertainty in gas supply and demand and the information likely to be available to each party.

Ideally, a model of the contracting process would predict the set of agreed-upon contract terms on the basis of these characteristics. Analysis of how the parties are affected by and respond to particular terms is an intermediate step in the analysis needed to determine the outcomes that those particular terms will produce. The acceptability of a set of contract terms depends on how the actions they produce are evaluated by the negotiating parties.

This section proceeds in several steps to illustrate the application of principal-agent theory to natural gas contacts. First, sources of uncertainty and information sets characteristic of gas markets are described. Then the risk allocations and incentives created by definite price escalators and take or pay provisions are compared with risks and incentives created by some alternatives. Finally, issues involved in applying the principal-agent approach to the natural gas contracting process are described.

Sources of Uncertainty

The risks that concern producers and pipelines are income, or more precisely profit, risks. Thus price uncertainty and sales uncertainty must be considered simultaneously to ascertain the risks involved.

The source of uncertainty affects the relation between price and quantity risks (Newbery and Stiglitz, 1981). A shift in the demand curve for gas will cause prices and quantities to fall, thus reducing all producers' income. A fall in supply by one producer will increase both the price received and the quantity sold by all others, though in aggregate the quantity sold will fall. For example, if gas demand has an elasticity of -1, fluctuations in supply will cause exactly compensating fluctuations in prices, so that total producers' income will be relatively stable.

When prices were regulated, market (as opposed to regulatory) risks affected income only through their effects on quantities, and persistent gas shortages ensured that all available supplies were taken. Producers faced no market risk, but also had no incentive to bring forth output commensurate with the value consumers place on gas. Pipelines also faced relatively little risk, in that they could easily sell all gas they could obtain, but did see opportunities forgone in random fashion because of their inability to obtain gas to meet demand.

Deregulation of field prices increases income risks of both parties, because prices become uncertain and because shifts in demand will actually affect what can be sold rather than merely altering the level of curtailments. The interaction between price and quantity changes must be examined to determine the absolute change in income risk. In particular, pricing and delivery terms of contracts must be analyzed together to determine income risks facing each party.[3]

Information

To determine what variables can trigger contractual obligations, it is necessary to specify what information is available about conditions in the market, costs of gas production, and actions taken by other parties.

In applications to the natural gas producer/pipeline contract, it is realistic to assume that the pipeline has some information about final de-

3. This issue also arises in distributor-pipeline arrangements. Analysis of those arrangements also should address the curtailment risks inherent in relations between distributors and final consumers.

mand not available to the producer, and that the producer has information about field supply not available to the pipeline. The pipeline's actions will affect sales and burner tip prices, and the producer's will affect available supplies and costs.

Keying take or pay provisions to deliverability, as is done in existing contracts, requires some monitoring by pipelines who must be able to verify how much gas could be produced. But this is less information than is required to write a forcing contract, which includes the short-run marginal cost of production and the nature of longer run investments that affect deliverability. Thus the best contracts that can be written must leave producers considerable flexibility to determine the rate at which dedicated reserves are drawn down. A contract covering all possible contingencies would be impossibly complex, even if adequate information were available. The optimal production rate for gas at any point depends on the demand curve facing the pipeline and the cost of gas production. Both are uncertain. Gas demand varies with many factors, including the weather, and the cost of gas production includes both a short-run operating cost and a user cost. Operating cost may be difficult to predict because reservoir mechanisms are uncertain and because even aboveground equipment can fail, and user cost is uncertain because it depends on future prices and the effect of current production on future costs and capacity.[4]

A pipeline can also take actions that affect returns to the producer but on which the producer has no direct information. These include contracts the pipeline signs and other ways it deals with its customers and the regulatory process.

Comparison of Contract Terms

Delivery terms, such as minimum take provisions, allocate quantity risks *among* producers as well as between producers and pipelines. Producers with more stringent delivery clauses face less risk than those with less stringent clauses, and an increase in the stringency of some contracts increases the risks faced by remaining producers. These market equilibrium effects are largely unanalyzed in studies of contracting. Price provisions have the same effect indirectly because price changes motivate pipelines to change their sources of gas and, when moved through to consumers, affect final demand. Price increases will reduce demand for total production, and delivery provisions will allocate the decrease among producers. Thus a producer with high price escalators and a high take or pay requirement will acquire a disproportionate share of the market when demand falls. A more uniform allocation of risk among producers would probably improve market efficiency.

A contract with definite price escalators and stringent take or pay provisions shields the producer from most market risk. It does not protect from supply risk—drainage by others, unanticipated pressure drop, or equipment failures—but these are individual risks and even a producer of moderate size may face small risks when income is averaged across several properties. A fixed payment, independent of production, would insulate the producer from all risks, and is never made because it removes all incentives to produce gas.

For both producers and pipelines there is a tradeoff between price and quantity terms. The most stable cash flow is created by specifying a definite price and a high take or pay obligation. The more risk averse a producer is, the lower will be the price that the producer will accept in return for stability of revenues. A less risk-averse producer willing to tolerate the risk of being shut in would command a higher price. (An aversion to being shut in may also be created by the threat of drainage, and thus create a preference for high take or pay levels independent of attitudes toward risk.) Diversity in contracting, with more or less stability in producers' revenues depending on the relative attitude toward risk held by the parties, would allow a matching of interest.[5] Where the tradeoff between the level and the stability of expected

4. If demand depends on a vector of random variables contained in some set X and production cost on another vector of variables contained in some set Y, an optimal contract would have to specify a production rate for every combination of values in the Cartesian production $X \times Y$.

5. We are grateful to George Hall for suggesting these points.

revenue lies in any individual contract depends also on how incentives are affected by the agreed-upon level of protection. Definite pricing provisions and high take or pay levels may not motivate efficient production decisions under conditions of changing demand. This motivation can come only from pricing or offtake provisions tied to movements in the final demand schedule. For example, if a fixed price or escalation formula is chosen, producers will no longer face uncertain prices but they also will not have an incentive to adjust production rates to match shifts in demand. Even if pipelines are allowed to vary their purchases freely, they will not have the information or incentive to identify the least costly (as opposed to lowest priced) sources of additional supply, or to match the marginal cost of production from a given source to the demand price. Likewise, changes in the cost schedule of gas producers will not be translated into changes in the cost of gas to pipelines. Even if producers can reduce their deliveries, there will be no market mechanism to bring forth the least costly supplies to meet demand or to reduce demand until willingness to pay for gas equals marginal cost.

Rigid price and offtake provisions can produce excess supply or demand for gas at any point. These effects increase the income risk facing pipelines, who may forgo profits in good markets if additional supplies cannot be obtained and will definitely suffer from paying fixed costs and penalties in poor markets. Whether any profit risk results depends on how the pipeline is regulated.

Similar effects can be seen in transactions between pipelines and distributors. Definite price terms and minimum take requirements may also insulate the pipeline from market risks which do not disappear but are shifted to the distributor. Whoever bears these risks is likely to concentrate on marketing gas efficiently.

If the pipeline makes a predetermined payment for a fixed amount of gas, it will face risks that provide motivation to sell gas in the most profitable permissible way and thus will be willing to offer a larger fixed payment than under any other arrangement. But this payment method will expose the *pipeline* to all risks of changes in demand.

Issues in Applying the Principal-Agent Approach

While analysis of the principal-agent problem sheds some light on natural gas contracts, transactions in the industry are more symmetric than assumed in the theory of agency. It is not always clear which party to the transaction should be considered a principal and which should be considered an agent. The pipeline is clearly the agent in transactions with distribution companies and industrial customers. But a pipeline is an agent selling gas for the producer as much as the producer is an agent managing gas reserves for the pipeline. Moral hazard problems may exist on both sides.

In its simplest form, optimality in a principal-agent relation involves choice of that fee schedule, contingent on outcomes, which results in payments to and performance by the agent that maximize the principal's expected utility. Acceptance by the agent is assured by constraining the agent's utility to exceed some level determined by the agent's bargaining power (or the utility he could achieve in some alternative employment) (Shavell, 1979). Different optimal fee schedules may be chosen, depending on the level at which the agent's utility is set. If it is assumed that contracting parties will agree on a Pareto-optimal contract, the results of principal-agent theory can be used to predict what kinds of fee schedules (or more general provisions) will appear in natural gas contracts. The prediction of principal-agent theory is that observed contracts will be members of some set of fee schedules such that if any one is adopted, it will be impossible to make one party better off without making the other worse off. Choice of a particular fee schedule in this set will then depend on the relative bargaining power of the two parties.

There is a basic asymmetry in the relationship between principal and agent, because only the agent takes actions that affect outcomes, and only the relation between the agent's effort and the outcome is uncertain. This asymmetry may not be present in relations between gas producers and pipelines. The value of gas to the pipeline is uncertain and this demand uncertainty is more important by far than uncertainty in the relation between producer's effort and gas output. Demand uncertainty may arise either from conditions in final markets for gas or from actions of regulators. To represent this uncertainty, it could be assumed that the pipeline's utility depends on some function of gas output and a random variable representing the state of demand. Sharing the risks introduced by demand uncertainty would then be an additional consideration in constructing an optimal fee schedule.

A pipeline may also take actions which affect the value of gas production, based on information not available to the producer. Coordination of actions in situations where all parties have identical preferences is the subject of the "theory of teams." Recent attempts have been made to analyze coordination of actions with conflicting preferences and imperfect information. This research merges the theory of teams with principal-agent theory (see Green, 1982, for references).

Extensions of this theory are required to identify rigorously the types of contract provisions that would be optimal with the types of uncertainty and possibilities of joint action that seem typical of gas markets. However, the conclusion that risk sharing is necessary is likely to stand. What extension of the theory may provide is a means of characterizing contract provisions more complex than the simple fee schedules found in the existing literature. Such an extension is needed if the theory is to provide any detailed understanding of how improvements might be made in existing contracts, whose complex provisions appear to be designed to deal with moral hazards on both sides of the transaction.

Implications

The conflict between risk sharing and incentives suggests some likely consequences of field price deregulation for long-term contracts. Prior to NGPA, uncertainty about prices for producers and pipelines was minimal and came from events such as changes in the nationwide uniform rate or deregulation of the entire industry. Thus risk sharing was a smaller problem in constraining contract terms. With the uncertainty of gas demand and prices unmasked, long-term contracts must become more complex or they will create incentives that lead to less efficient resource allocation. Either effect (or both) constitutes an increase in cost of contracting that is likely to push the industry toward either greater integration or greater reliance on spot sales.

The presumption that increasing uncertainty requires changes in the nature of long-term contracts is amply confirmed by recent developments in natural gas markets. With an unexpected drop in real oil prices and a serious recession, natural gas demand has fallen. Simultaneously, pricing provisions of contracts have stimulated gas production and take or pay provisions have required pipelines to buy more gas than they can sell. The nature of existing contracts, in short, has hindered the transmission of falling demand signals to producers and thus the incentive to reduce production.

The principal-agent approach leads to the conclusion that greater uncertainty will make contracts a more costly and less efficient means of governing transactions. However, the direct application of that approach to optimal incentive contracts does not provide definite answers to the question of how gas contracts will or should be designed. For example, the conclusion that an optimal contract should include a fixed fee and a sharing rule is derived from a model in which information and control over outcomes are asymmetrically distributed. Relationships between gas producers and pipelines are more symmetric in both these dimensions, and extension of the principal-agent paradigm to deal with this symmetry case is required if definite conclusions about optimal gas contracts are to be rigorously derived.

NPGA and New Market Institutions

None of the transactional arrangements discussed heretofore promise optimal coordination of vertically related activities. Vertical integration appears to achieve an optimal result through central direction, but its virtues are tempered by costs and difficulties of internal organization and by potential effects on horizontal structure. Contracting fails to provide adequate incentives for making production and other decisions in a way that will meet final demand most efficiently, and spot transactions could be undesirable because of the way they load risks onto a single party even if no problems of market power created by "relation specific" investments existed. Fundamentally, spot market transactions fail to achieve optimality because of the lack of adequate risk markets, so that natural gas transactions must perform two inconsistent functions: coordinating production and transportation functions and sharing risk among the parties. If new market institutions could be developed to shift risks separately from the transaction between producers and pipelines, the conflict could be reduced.[6]

Futures markets are an example of such a new institution. Futures contracts in essence bring a third party into the transaction, who may find bearing risks associated with price changes more attractive than do pipelines or producers. The third party could be one who finds that income from other sources is negatively correlated with the price of natural gas, so that natural gas futures have a hedging or insurance value. Introducing such a third party makes it possible to set optimal incentives for pipelines and producers without regard to their immediate effects on risk. Risks can be reduced through secondary trading on futures markets.

Efficient production of gas under contract requires that the end-use price of gas be taken into account by the producer, who can adjust production to match price and marginal cost. A contract might, for example, call for sales at "spot" prices determined in a related market. This formulation of a contract places the entire burden of price uncertainty on the producer, as opposed to a "forward" contract for delivery of a specific quantity at a fixed price, which places the burden on the purchaser. There is an analogy to agricultural markets in which producers awaiting harvest or holding stocks of grain face uncertain spot prices. Forward contracts, locking in a price for a specified quantity of grain at a specified date, eliminate price uncertainty. Alternatively, a farmer can to some extent "hedge" by selling grain futures contracts. In this case the farmer will eventually sell grain on the spot market and simultaneously buy, expiring futures contracts to cancel out his position. If the price of an expiring futures contract is identical to the spot price, all price uncertainty will be eliminated by the transaction.

In agricultural markets expiring futures contracts are not priced identically with spot transactions, and the residual uncertainty is termed "basis risk." Forward contracts differ from futures in that actual delivery of a specific quantity at a specific price is required. Such contracts eliminate "basis risk," but are possible only after finding a purchaser who demands the specific packet of gas involved. Current gas contracts are forward contracts extending over a sequence of transactions and with contingent prices. The durability of gas reserves makes the relatively short-term forward contracts used in agricultural markets of little use in reducing producer's risk.

This combination of futures contracts with spot sales provides an incentive for efficient production not present with forward contracts. With a forward contract at the expected price, production will take place up to the point at which marginal cost equals the expected price. With spot sales hedged by futures, the producer can obtain a spot price greater or less than the expected price. If the spot price is greater, the producer can increase production, earning greater profits than would be earned at the quantity optimal with the expected price. Thus, an increase in production to meet a higher demand is motivated by a combination of spot and futures sale, but not by rigid forward contracting. If fully hedged, the farmer would take a loss on futures equal to the difference between the expected price and the spot price on the fixed quantity of futures (equal to optimal production at the expected price). Similarly, if the spot price turns out to be less than the price at which futures were sold, the producer will cut back production and make up required deliveries by spot purchases, making a profit on futures equal to the difference between the expected price and the spot price.

Moreover, the producer will earn profits on his additional sales when the spot price is high, and avoid production costs when the spot price is low. Thus selling spot, hedged by futures, gives profits higher than could be achieved with

6. For more on the general problems of markets under uncertainty and means of improving outcomes, see Newbery and Stiglitz (1981).

fixed production to satisfy a forward contract in every state of the world. This conclusion is illustrated in figure 4-1. If exactly Q_0 is produced, the cost of retiring futures contracts is exactly equal to revenues from spot sales, so that the producer ends up with gross revenues equal to original revenues from selling futures Q_0E_0 (P) and with profits equal to the area $D + E$. If Q is produced, and sold at a price P, additional revenues equal to $P(Q - Q_0)$ and additional profits of A are obtained. If Q' is produced and sold at price P', gross revenues are reduced by P' $(Q_0 - Q')$, but production costs are reduced by $B + P'$ $(Q_0 - Q')$, so that profits are increased by B.[7]

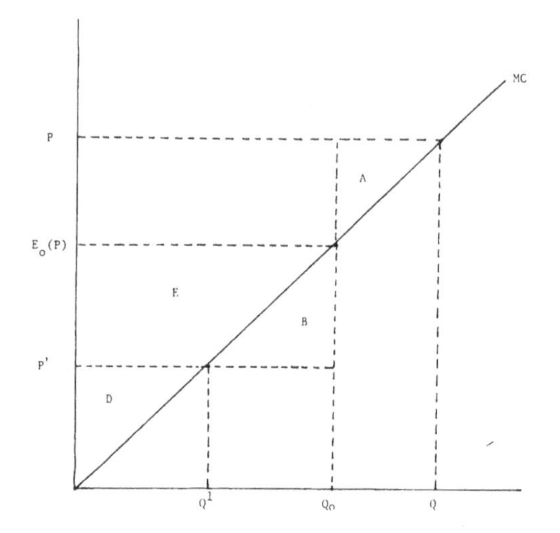

Figure 4-1. Profits from Hedged Spot Sales

If reserves were not dedicated, forward and spot transactions would be identical. More than the contracted amount could be produced and sold on the spot market if the spot price were high, and less could be produced and the balance made up from the spot market if the spot price were low.

However, dedicating reserves eliminates a source of moral hazard that could otherwise be dealt with only through more complex contractual terms and monitoring the physical condition of a reservoir. The integrity of the gas reservoir must be protected from actions by the producer if any discretion on delivery is to be allowed the producer. If the producer is allowed to make spot sales when prices above the contracted level are available, the contractual purchaser may find future availability at contracted prices reduced. Monitoring the producer's actions directly and assessing their impacts on future contract deliveries would be necessary if reserves were not dedicated. Such monitoring is possible but costly, so that reserve dedication can be seen as reducing transaction costs.

Despite the virtues of the combination of spot and futures markets, there are obstacles to the development of either or both. (It should be noted that a true spot market, or other freely accessible means of making delivery to an open contract, is a prerequisite to the development of a futures market.)

One obstacle is that certification and bond covenant requirements of assured supply might not be satisfied by reliance on spot purchases. The more pervasive and competitive spot markets become, and the greater the availability of standardized "futures" contracts, the less this would be a problem.

Another potentially important obstacle to spot sales becoming the rule is the specialized nature of production and gathering equipment. A producer who loses a purchaser after developing a field may be in a position of making distress sales to any future customer; and a pipeline that has invested in gathering lines also faces substantial losses if an historical producer were to decide to sell elsewhere. These are exactly the conditions identified by Williamson and others as promoting opportunism and making spot transactions potentially costly. (Franchise monopoly and common carrier status, discussed in chapter 5, also could be imposed on pipelines, thus enabling spot transactions between producers, large users, and retail distribution companies to grow.) The magnitude of the costs of ending a relationship between pipeline and producer will be critical to the viability of a spot market.

Spot purchases can take place without organized commodity markets, although such markets would economize on the cost of information for small purchasers now relying on the brokerage function of pipelines. However, a spot mar-

7. This result is a special case of the general proposition, demonstrated first by Oi (1961), that expected profits are higher when a producer faces uncertain prices than they are otherwise.

ket without a futures market could put an unacceptable risk of price changes on the producers. A futures market allows hedging to reduce this risk only if gas is delivered at standard locations and prices. Such standardization of spot transactions is necessary if there is to be adequate correlation between spot prices and the price of an expiring futures contract on any given date.

Because of substantial transportation costs and varying distances between fields and markets, a uniform nationwide field price is unlikely. Thus separate futures prices for delivery at major pipeline intakes would be required. Each such market would have to handle enough business volume to supply hedging needs and to avoid manipulation.

Regulators of futures trading would also have to be particularly vigilant (Houthakker, 1959). Because of the substantial penalties for failure to deliver on an open futures contract, corners and squeezes are methods of market manipulation that must be guarded against. (In a corner, one party buys up all contracts for delivery on some date, and can extract high prices on the closing date from any trader who sold futures short; in a squeeze, the corner is extended to tie up available spot sales, so that satisfaction of expiring futures contracts is impossible.) Position limits are the usual methods of preventing manipulation, but they further increase the size of a futures market needed to satisfy large purchasers and sellers. Corners and squeezes may be particularly troublesome in gas futures markets because of the possibility of pipeline bottlenecks interfering with physical delivery of gas to satisfy particular expiring contracts.

These considerations suggest that if spot sales are to occur at an adequate scale, virtually all transactions may have to be on the spot market. The transition to such a situation may be particularly difficult if no intermediate scale of spot transactions is viable.

Impacts of Regulatory Constraints on Transactional Arrangements

The general consequences of regulation for the behavior of firms—primarily with regard to input substitution—are discussed elsewhere in this study. Another effect of regulation, however, is on the choice of transactional arrangements. This issue has scarcely been treated in the transactions cost literature reviewed above, but is likely to be particularly important in natural gas where institutional and regulatory factors overlay economic factors. There are two related questions. First, does utility-type regulation, prevalent in downstream natural gas sectors, influence transactional arrangements in an industry comprised of distinct vertical markets? And second, will deregulation of one market change the cost of conducting transactions even when markets vertically related to it remain under regulation?

On the first question, the literature is sparse. However, Peles and Sheshinski (1976) offer an insightful analysis of the effects of regulation on horizontal integration. They construct a simple Averch-Johnson model of a firm that jointly produces a multitude of products with unrelated demands. They show that with a constraint on rate of return, horizontal integration—for example, common ownership of utilities serving contiguous but separate geographic areas—increases total company profits by allowing the composition of outputs to change in accordance with relative profit opportunities in the various markets. While the Peles-Sheshinski model does not directly apply to vertically related markets—with firms producing multiple products but with (vertically) related demands—a modification of their model is possible. Of course, because the framework is applicable only to markets under rate-of-return regulation, it is not suitable for analyzing behavioral responses involving firms under first-price regulation (that is, it could not be used to look at transactional arrangements between natural gas wellhead production and transmission or distribution companies), but could be useful for analyzing incentives for transactional arrangements between the pipeline and distribution segments. It could be applied to horizontal structural changes that may come about as a result of upstream deregulation of natural gas, as discussed in chapter 3.

Regulatory provisions permitting automatic adjustment of changes in input prices, such as purchased gas adjustment clauses for gas transmission and distribution companies, are another facet of contemporary rate-of-return regulation that may affect the structure of vertical transactional arrangements. Fuel adjustment mechanisms may lessen the incentive of distribution companies to integrate into production to secure more certain supply. Even with fuel adjustment mechanisms, pipelines have an interest in expanding markets (and hence their rate base) and are unlikely to be completely indifferent to increases in input costs. The regulatory lag involved in formal rate cases creates a stronger incentive to minimize cost, but also increases the transaction cost associated with changing prices. To the extent that a pipeline has discretion about curtailing customers, the lack of a fuel adjustment mechanism may make supplies to the pipeline's customers less certain. Thus fuel adjustment mechanisms may make downstream distributors more certain about supplies under circumstances of uncertain costs. In this way, fuel adjustment mechanisms may directly affect the nature of transactional arrangements by reducing the drive toward vertical integration to obtain greater assurance of supplies than is available under alternative arrangements.

Finally, regulatory status may lead to certain types of transactional arrangements. For example, most gas pipelines historically have operated as private carriers, taking ownership of the commodity transported. There are alternative regulatory regimes. In "contract carriage," private parties contract for a pipeline's transportation services and the pipeline does not play a buyer/reseller role. In "common carriage," pipelines may be *required* to transport gas for any producer or distributor at a regulated tariff. In contrast, private carriage gives pipelines an interest in assuring future supplies to customers because this will reduce the cost of having excess capacity. It also gives them an incentive to exercise some degree of control over input prices in order to maintain or increase market share and hence the rate base. Thus, to the extent there is a change in the regulatory status of pipelines toward contract and common carriage, there will be fewer incentives for vertical integration.

Changes in vertical transactional arrangements as a result of NGPA, while involving the downstream sector, could originate primarily from incentives created at the upstream level. For example, producers are also concerned about reducing transactions costs and uncertainty about demand and output. However, because they are not subject to utility-type regulation, and under NGPA increasingly subject to less regulation of any form, their response is likely to be different than that of transmission and distribution companies. Moreover, although the private carriage regulatory status of pipelines creates incentives for downstream vertical integration on the part of producers, a change toward contract or common carriage may blunt those incentives.

The effects of deregulation in one market on transactions costs in vertically related industries still under regulation is a different issue. The existence of a regulation differential may produce fundamental changes in transactional arrangements. As argued above, opportunistic behavior on the part of contractual parties is likely to create incentives for noncontractual arrangements. However, as pointed out by Goldberg (1976a and b), government regulation may prevent opportunistic behavior. Thus relaxation of regulation could give rise to an increase in the use of noncontractual arrangements. It is on these grounds, as Klein, Crawford, and Alchian (1978) argue, that the difference in vertical integration between the oil and natural gas industries may have resulted because of less effective price regulation of oil. In other words, more complete government regulation of gas prices may prevent opportunistic behavior by gas pipelines, and thereby serve as an alternative to vertical integration. Two empirical observations lend support to this hypothesis. First, there is a greater degree of vertical integration in the *intrastate* natural gas industry than its *interstate* counterpart. Second, as relaxation of field price regulation has become more and more certain, along with the passage of NGPA, there has been an increasing amount of vertical integration. (A third piece of evidence has already been mentioned: in the early history of the industry, prior to the Natural Gas Act, vertical integration was much more pronounced.)

Policy Implications of Transactional Arrangements

That many industries throughout the economy are characterized by nonmarket transactional arrangements suggests that the "invisible hand" of the market is not always the most cost-effective way of maximizing economic welfare. The transactional arrangement literature shows there are many circumstances in which contracting or integration can help correct this market failure. However, there is no guarantee that transactional arrangements adopted by producers, pipelines, and distributors will be in the general interest.

Uncertainty created by field price deregulation clearly requires changes in the nature of long-term gas contracts. Take or pay and price escalation clauses in contracts between producers and pipelines and minimum bill provisions in contracts between pipelines and distributors serve to insulate a given segment from changes in price and quantity that are occurring in other segments. More flexible provisions are needed to create incentives for producers to adjust their output to meet fluctuations in demand as well as incentives for pipelines to select the least-cost mix of supplies. The degree to which such incentives place a disproportionate share of risks on one party to the contract can be moderated by different types of pricing, such as combination of a fixed fee for storing gas during periods of slack demand with payment for gas produced, or by introduction of market institutions that offer new possibilities for shifting risks, such as futures markets.

As Carlton (1979) and Okun (1981) have shown, society may bear costs associated with market rigidities created by current contracting practices. If contract provisions do not allow for smooth transmission of changes in demand and supply between vertical stages of the industry, markets may not reach equilibrium. This disequilibrium can create social costs that extend beyond the losses felt by the parties to the contract. As a result, there may be a public interest in moving contracts toward greater flexibility and provisions of more effective incentives to producers.

The increased costs of contracting in an uncertain world may also produce a tendency toward greater vertical integration that could also have a significant impact on economic welfare. As Carlton (1979) and others have shown, the private gain from vertical integration may be more than the social gain even if such an arrangement economizes on transaction costs. The literature on vertical integration, particularly Bork (1954), suggests that integration by firms with sufficient horizontal dominance—that is, high market shares—can result in a loss of economic welfare. More subtle effects of the interaction between vertical integration and horizontal features of markets, as argued by Broadman (1981), could also lead to a divergence between private and social welfare. Competition *within* natural gas markets as well as *across* markets could be affected by such a structural change. Private incentives for transfer pricing and market cross-subsidization under a vertically integrated structure need not bring about socially desirable outcomes, and policies that are effective in ameliorating effects of vertical integration in other industries may be inappropriate to the regulatory and institutional system prevalent in natural gas.

Research Needs

Because of their central and continuing importance in gas markets, contractual arrangements should be given high priority in research. Basic theoretical research is required to expand the principal-agent model to include uncertainty in the needs of the principal as well as in the effects of the agent's action and to allow for action by the principal that affects the utility of both parties. This model can be applied to natural gas contracts in two steps:

Characterizing the risks and incentives created by specific contract terms and predicting behavior of each party in response to those terms

Representing contract terms as a matter of negotiation between the parties and finding equilibrium and optimal outcomes of that negotiation process

The result would be a model able to predict provisions of contracts that would be adopted under various conditions, which could be tested against historical experience in interstate and intrastate markets before and after deregulation. An evaluation of the results of the contracting process should take into account potential divergencies between the interests of parties to a particular contract and those of the economy as a whole.

Analyzing the mix of transactional arrangements requires comparing optimal long-term contracts with alternative transactional arrangements. One line of research should deal with the prospects for the emergence of new market-based transactional arrangements. Natural gas will continue to be produced inefficiently even with field deregulation if contractual arrangements continue to insulate suppliers from changes in gas demand or allocate those changes among suppliers without regard to their costs of production. In general, contract terms provide incentives for effort and divide risks among the parties. "Netback pricing," in which the pipeline subtracts transportation retail price costs from the producer's, creates incentives for efficient gas production but places all risks of price change on the producer. On average, higher prices will be needed to compensate producers for bearing such risks than would be required with fixed escalators that place price risks with the pipeline. If alternative sources of insurance or hedging, such as futures markets, could be developed, efficiency in production and in allocation of risk would be reconciled. However, futures markets cannot exist without spot markets, and the transportation costs and specialized capital characteristic of the gas market may make a pervasive spot market impossible.

Research should begin by examining capital equipment involved in the production and gathering of natural gas, and the nature of competition among producers to assess the importance of economic barriers to spot transactions. Whether spot markets will establish sufficiently uniform prices and whether the proportion of futures to spot transactions will be enough to provide adequate hedging must be determined. Agricultural or other commodity markets can serve as a yardstick to the extent they display analogous technical conditions and sources of uncertainty. Actual contract terms should be evaluated according to their risk-sharing properties and their ability to allocate gas efficiently. Rough numerical estimates of social surplus losses from rigid contract terms can be made.

Analysis of the prospects for increased vertical integration requires extended theoretical research on the choice between market and nonmarket processes for coordination of vertically related activities. The approaches of Arrow and Carlton to informational and cost considerations in vertical integration have particularly strong potential for natural gas markets.

Some criterion must also be used to assess the implications of changes in transactional arrangements for economic welfare. The effects of regulation on the incentives of upstream and downstream parties will be an important factor in this overall assessment.

5

Upstream Deregulation and the Behavior of Downstream Firms Under Continued Regulatory Restraint

Introduction

Continued regulation of natural gas pipelines and distribution companies by state and federal commissions creates constraints and incentives that must be considered in any analysis of the downstream effects of field price deregulation. How pipelines and distribution companies behave under their regulatory restraints will affect each link in the chain of transactions and activities that take natural gas from the field to ultimate consumers.

With deregulation of gas production, pipelines will face a freer market in which to bid for new gas supplies. They will also face a different environment in which they must make choices about building gathering lines, long-distance transmission capacity, and storage to match new patterns of demand that will emerge from higher delivered gas prices and the end of forced curtailments. Their choices regarding these "inputs" to the business of gas delivery will still be affected by the Federal Energy Regulatory Commission (FERC).

Pipelines deliver gas to other pipelines, to direct customers (mainly large, industrial customers), and to distribution companies who, in turn, sell gas at retail. How this gas is priced by pipelines to different classes of customers also will still be affected by FERC regulation (for interstate transmission) and by state regulatory authorities (for intrastate transmission), both of which set constraints on pipeline cost allocation and on pipeline rates of return. Thus, even pricing of nominally unregulated direct industrial sales by pipelines will continue to be affected by utility-type regulation, insofar as joint costs exist with regulated sales.

Retail distribution companies will also continue to be subject to rate-of-return regulation, but at the state and local level. Pricing of gas to ultimate residential and commercial, and to some extent, industrial consumers, is shaped by regulation of these companies. Regulation of distribution companies also affects their derived demand for gas from pipelines. By its nature, however, such regulation is more difficult to codify and analyze in general terms because of the variety of approaches taken by different state commissions.

By applying the literature on the economics of regulation to stylized facts about the gas industry, an analytical framework can be devised which can help predict and evaluate the effects of continued downstream regulation on the input choices and pricing strategies of pipelines and distribution companies in the new regime of upstream deregulation. To this end, although these two categories—input choices and output pricing—in principle, cover the entire spectrum of downstream activities, they must be further subdivided in order to explore fully the patterns of regulatory influence on downstream behavior.

Four specific subjects deserve close attention. First, precisely what is the process by which regulation influences input and output choices of firms? Analysis of that interaction will shed light on the elements that determine whether, under partial deregulation, prices of unregulated gas will be bid up to unjustified levels. Among other things, it could also show the extent to which regulation creates a bias toward an inefficient mix of line pipe and storage capacity.

Second, what distinguishes the structure of prices that will be chosen by (1) an unregulated monopolist, (2) a regulated monopolist, and (3) a public enterprise whose objectives are maximization of consumer welfare? By applying the literature on price discrimination, it is possible to analyze the practice of charging different gas customer classes prices that differ on the basis of demand elasticities as well as the practice of setting different gas prices for peak and off-peak demand. Although regulatory commissions place some constraints on such pricing strategies, gas transmission and distribution companies retain considerable discretion. Thus predicting how pricing patterns will evolve as field prices rise and as pipelines and distribution companies are freed to seek and serve new customers is not a trivial problem of simply applying regulatory formulas.

Third, how can the demand for reliability of gas service be both observed and met? Analysis of how prices should be set to meet stochastic demand should recognize the distinction between *ex ante* agreements on the degree of reliability and *ex post* delivery. For efficient functioning of gas markets, it is important to match willingness to pay for reliability with the cost of providing it. Such matching, however, is likely to be difficult in principle and may require particular patterns of price discrimination.

Finally, closely related to the issue of price discrimination is cross-subsidization across regulatory boundaries. What determines how regulated firms are likely to behave when they also operate in unregulated markets? Potential pricing distortions inimical to competition and to efficient resource allocation are of particular concern if unregulated, direct industrial gas sales become more important components of pipeline activity.

Each of these subjects is explored in the literature on the economics of regulation, but very few have been addressed in studies of the natural gas industry. Thus in this chapter we develop some general theoretical results and describe methods that have been used fruitfully in the study of other regulated industries. The theory suggests some preliminary insights into how regulated pipelines and distribution companies will behave after field markets are deregulated and how regulatory practice might be improved. The theory and the results of empirical studies in other regulated industries also provide a basis for recommending a strategy for research on downstream regulation.

Input and Output Choices

The first systematic analysis of how regulation affects input and output choices is by Averch and Johnson (1962). They formulate a model, hereafter the A-J model, in which a monopolist sets the price for its product subject to a regulatory constraint on its rate of return. The fundamental result of their analysis is that if the permitted rate of return set by the regulatory authority exceeds the current cost of capital, there is an incentive to expand the rate base. In short, rate-of-return regulation has the effect of reducing the cost of capital below its shadow price. Thus the regulated monopolist will choose a ratio of capital to labor inputs that is larger than the cost-minimizing level.

The A-J hypothesis can be demonstrated geometrically, as in figure 5-1. Suppose to produce a given level of output, Q_1, the utility faces long-

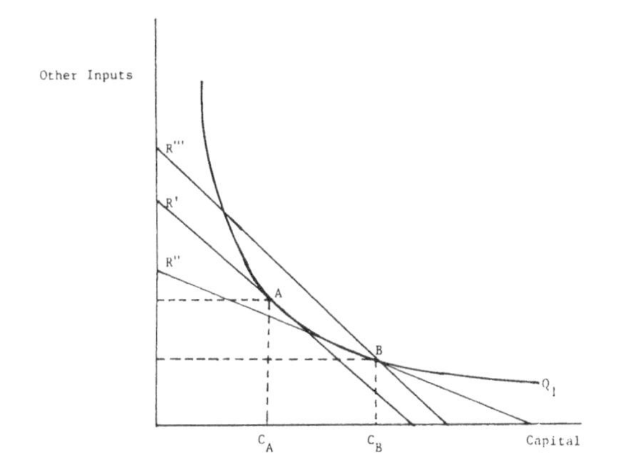

Figure 5-1. Input Choice Under Regulatory Constraint

run substitution choices of capital and other inputs traced out by the curve labelled Q_1. If the relative market cost of capital is depicted by isocost line R', output of Q_1 should be produced at point A, using C_A units of capital. But because regulators set the allowed rate of return above the market cost of capital, the utility perceives a lower private cost of capital—one consonant with the isocost line R''. Thus, to produce the same Q_1 units of output, it will operate at point B and use $C_B > C_A$ units of capital; capital becomes relatively cheaper than other inputs to produce the same level of output. The cost to society of producing at B, however, is greater than to the individual firm, since R'''—which reflects the relative social cost, or shadow price, of capital—is not tangent at B and above the isocost schedule R'.

Numerous variants of the A-J model have been developed. Analysis has proceeded along three major lines:

Formulating more precisely what the regulator does: what actions of the regulated firm are controlled and how that control is exercised

Examining the impacts of regulation under changing economic conditions affecting the regulated firm

Exploring a wider range of decisions by the regulated firm and how those decisions are influenced by regulation

Bailey and Malone (1970), for example, look at four plausible formulations of the regulatory constraint: rate-of-return, total profit, profit per unit of sales, and fixed markup. They show that overcapitalization occurs only under rate-of-return constraint, and that distortions in the purchase of other variable inputs also depend on the nature of the constraint. Bailey and Coleman (1971) introduce regulatory lag—the length of time between the filing of a request for a rate increase and the issuance of a decision regarding the request—into an A-J model. They find that as the regulatory lag becomes longer, the degree of distortion becomes smaller.

In a comprehensive survey of the literature following Averch and Johnson, Baumol and Klevorick (1970) demonstrate that different results are obtained when it is assumed that commissions set price schedules, rather than leaving that to the regulated monopolist—which is the way the A-J model formulates the problem. They concur with the conclusions that lags in adjusting prices reduce the bias toward overcapitalization, and develop a new model of innovation by a regulated firm to show that regulatory lag also provides temporary profits that give a stimulus to innovation.

Joskow made the first radical departure from the A-J tradition by pointing out that rate reviews do not occur randomly or at fixed intervals, but rather can be brought about by the regulated firm or by other parties when profits appear excessively high or excessively low. Thus regulatory lag becomes an endogenous variable, subject to change by actions that alter the regulated firm's profits. There is an incentive to produce inefficiently only when profits approach a level that would trigger a rate review. At other times, minimizing cost means maximizing profits, for all cost savings are retained by the firm.

In a series of three papers, Joskow (1972, 1973, 1974) laid out a new path for analyzing the behavior of regulatory commissions and how their decisions affect regulated firms. In the earliest article, he formulates a model for predicting how commissions will determine a "fair" rate of return in a formal hearing. Consideration is not given to either the determination of the rate base preceding the rate-of-return decision or of the rate structure decision which follows. He concludes that commissions behave in a consistent fashion despite the lack of explicit rules for calculating a fair rate of return. A finding of considerable methodological significance is that there is substantial interaction between the behavior of commissions and firms. Given this interaction, characterizing firm behavior as one of maximizing profits subject to an exogenous rate-of-return constraint, as is done in conventional A-J models, is not correct.

In the second article, Joskow (1973) determines which variables are important in prompting a regulated firm to seek a rate change, thus filling in one of the givens in his 1972 article. Decisions by firms to petition for a rate review are modelled as a function of expected commission response. The model explains why utilities have, at times, "swallowed" large cost increases without requesting a rate increase and at other times have requested rate increases in response to relatively small cost increases. It assumes that firms make decisions on the basis of a number of relatively limited, short-range "rules of thumb" which indirectly reflect long-term goals. Joskow argues that this characterization is especially relevant to regulated firms not facing actual or potential competitors. The empirical results are that the decision to file for a price *increase* depends on the firm's earnings, on the level of interest coverage, and on "a variable which measures prior expectations of success in the hearing room." Decisions to file for a price *decrease* are also found to be determined by earnings and "a variable which measures the expectations of 'forced' regulation" (Joskow, 1973, p. 118).

Joskow emphasizes that decisions to file for a rate change—often rate increases—increasingly are viewed as risky by utilities for three reasons. First, until about 1967 it was possible that in the course of reviewing a requested rate increase, a commission would decrease rates because average costs had declined. Second, because it can take commissions several months to reach decisions and because no additional requests can be filed in the interim, a decision to file precludes the possibility of filing again for at least several months even if costs begin to change. That is, lengthier regulatory lags increased the risks associated with filing. Third, formal hearings often involve confrontations with intervenors and extensive financial scrutiny. Altogether, these factors resulted in price rigidities, with firms seeming to tolerate earnings fluctuations without adjusting prices accordingly.

An important insight of this analysis is that there is an upper and lower bound on the rate of return, the effects of which are not symmetric. Profit increases resulting from decreasing costs while prices are fixed are less constrained than those due to a price increase. Usually, price decreases were initiated by the utility to forestall possibly more severe regulatory action. As a result, there is likely to be less overcapitalization on the upper bound than the lower one; within these bounds firms are free to maximize profits by choosing inputs efficiently. When Joskow combines the effects of this rate-of-return boundary asymmetry with those of inflation, regulatory lag, and pricing based on historical costs, he concludes that regulated firms will generally behave as cost minimizers and not as depicted by Averch and Johnson.

In the third article Joskow (1974) addresses the choice of regulatory instruments. His thesis is that underlying economic and political conditions largely determine the performance of regulatory institutions and, in turn, the choice of regulatory techniques. He contends that the A-J model, in characterizing the typical regulatory agency as establishing and actively enforcing allowed rates of return, is inaccurate. In fact, Joskow argues, the process is essentially passive. Nominal prices, not rates of return, are what he observes commissions wishing to constrain. Moreover, he believes firms do not maximize profits *per se*. Rather, they act to maximize a set of financial parameters which they believe are consistent with long-run profits. And because the legislative directives for regulatory objectives and procedures are vague, there is a great deal more commission flexibility than portrayed by models in the A-J tradition. Finally, the rates of return generally allowed by a commission for firms that come under review tend to serve as benchmarks for all regulated firms—whether or not they are actually reviewed.

In general, then, Joskow maintains that the extent to which there are actual distortions in input choice appears to depend on the external economic and technical environment. Specifically, if nominal average costs are declining, the utility will minimize costs since it will be permitted to keep all profits earned at that price. But, if nominal average costs rise, the A-J type of biases may (but will not necessarily) appear. If an increase in efficiency can cut a firm's costs sufficiently to get it out of the regulatory review process, it will take that cost-minimizing route. The A-J theory does not allow for this "out," since the firm is presumed to be constantly under regulatory constraint.

Joskow also argues, following Simon (1959) and Cyert and March (1963), that organizations such as regulatory agencies seek to operate in stable and insulated environments and will try to do so by fashioning "rules of thumb" to standardize their decision-making process.

Empirical evidence is roughly consistent with Joskow's arguments. For example, throughout much of the postwar period there appears to be no systematic pattern to the number, duration, and timing of regulatory reviews for electric and gas utilities. With the advent of inflation in the 1970s, which would have increased the number and complexity of regulatory hearings and procedures, regulatory agencies began to adopt rules of thumb that include temporary rate increases, fuel adjustment mechanisms, use of future test years, and flattening in the rate schedule to reduce quantity discounts.

Overall, Joskow's representation of regulatory behavior implies a shifting of risk from the regulated firm to ratepayers. This shift comes about, argue Burness, Montgomery, and Quirk (1980), to the extent that regulation reduces the variance of the utility's rate-of-return. As a result, rate regulation causes a utility to show greater tolerance for risks than would the same firm in an unregulated environment.

This argument is extended by Isaac (1982), who develops a model of a utility subject to endogenous rate review and to a fuel adjustment mechanism (FAM). The FAM is found to provide an incentive for an inefficiently high ratio of fuel inputs to capital investment, and thus counters the Averch-Johnson bias. Most interestingly, Isaac develops a realistic representation of procedures of the California Public Utilities Commission (PUC), which incorporates an FAM and periodic rate reviews. He shows that while indeed regulation brings about a distinct allocation of input incentives, introduction of the FAM into the regulatory regime makes it impossible to describe the direction of the bias.

Surprisingly, there have been very few investigations of whether overcapitalization exists in the natural gas transmission industry. One exception is Callen's (1978) study. He formulates a model of a regulated pipeline in which capacity can be varied, and applies a constraint derived from the Atlantic Seaboard cost allocation formula to total revenues from jurisdictional sales. He postulates two alternative objectives for the pipeline, one of maximizing revenues and one of maximizing profits. Overcapitalization is defined in terms of the tradeoff between line-pipe capacity and horsepower capacity, assuming that horsepower inputs have a large variable cost component and line-pipe inputs none. The constrained revenue maximizing model is found to be the best predictor of input and output choices. The model predicts inefficiency in input choice, but the direction of bias cannot be stated unambiguously. In most cases, choice of a pipeline capacity/horsepower ratio lower than that which would minimize cost is predicted.

Callen also evaluates the performance of natural gas pipeline regulation. He does so by comparing the sum of producer's and consumer's surplus under optimal marginal cost pricing and under regulation. He finds that "(t)he effect of rate of return regulation and increasing output offsets the increased costs of input inefficiencies" (Callen, 1978, p. 319), and that net benefits under regulation are within 15 percent of their first-best optimum.

The central conclusion that emerges from this discussion of the standard A-J model is that it serves better as an example of how to characterize theoretically the effects of a regulatory constraint than as a direct source of useful insights about the natural gas industry. Its representation of the regulatory process is inaccurate and its concentration on the question of whether capital and variable inputs are combined in efficient proportions is misplaced.

The results that emerge from variants on the A-J model are, however, suggestive: the way in which the regulatory constraint is formulated is critical to the regulated firm's demand for inputs. Moreover, the approach originated by Joskow provides guidance on how to formulate a realistic representation of the regulatory process actually at work in natural gas markets. These conclusions lead directly to the issue of how the bidding for deregulated gas by regulated pipelines can be characterized.

Theory of Regulation and the Misdirected "Cushion"

The literature on input choices of a regulated firm can be applied to the question of whether unregulated gas prices will be bid up excessively. The proposition that partial deregulation will cause pipelines to bid the price of deregulated gas above the level at which field markets would clear under full deregulation is widely accepted. However, the argument depends on the assumption of a particular form of pipeline regulation, market conditions, and management objective.

Roughly, to conclude that deregulated prices will be bid above a market-clearing level, it must be assumed that pipeline tariffs are continuously adjusted to provide the pipeline with a fixed profit margin above average cost. Such regulation gives the pipeline an incentive to maximize throughput, and if deregulated gas is the only source of incremental supplies, the pipeline will purchase additional gas until average cost plus markup reduces demand to equal available supply. If demand for gas becomes perfectly elastic at some price, the quantity of gas sold will be increased to the point at which average cost plus markup equals that price. In the absence of regulation, deregulated gas would be purchased only until the marginal cost of additional gas reached the price at which gas could be sold.

This argument can be made more precise. Partial deregulation of field prices, as mandated by NGPA, implies that pipelines will face a schedule of prices for different quantities of gas. Gas supply contracts that did not allow for renegotiation or escalation to market levels would have the same effect. Some quantity of gas, that subject to continued price controls or contracted for at fixed prices, would be available at a price below the market. Additional supplies of gas would be available to the market at progressively higher prices. Ignoring for the moment the exhaustible character of the gas resource, the supply curve will be based on the increasing cost of supplying additional quantities of gas.

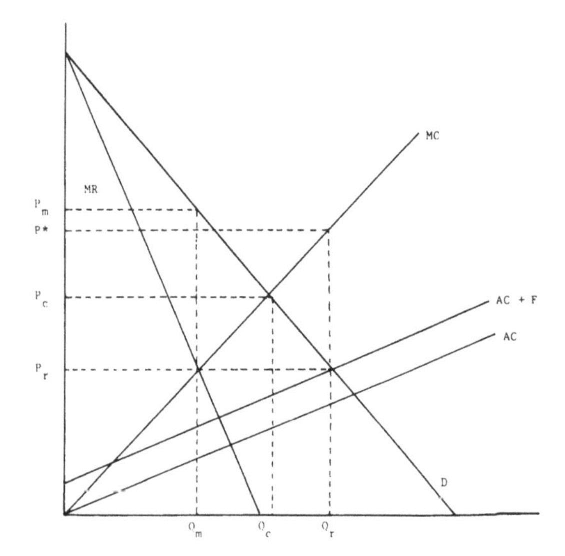

Figure 5-2. Monopoly, Competitive, and Regulated Prices and Quantities

If the regulator sets price at exactly average cost plus a fixed markup per therm of gas sold, the pipeline will be motivated to maximize sales. Even if no "obligation to serve" is enforced, the pipeline will supply exactly that quantity of gas at which the demand price equals average cost. This conclusion is illustrated in figure 5-2. The schedule of quantities of gas available to the pipeline is the marginal cost curve of the pipeline, *MC*. Average cost of gas, *AC*, is represented by a curve below *MC*. Average revenue is equal to average cost plus a fixed markup, *AC* + *F*. The demand schedule is *D*, and the associated marginal revenue curve *MR*.

A monopolist would choose to sell Q_m units of gas, at a price, P_m. A competitive firm would sell a larger quantity Q_c, at a lower price P_c. The profits of a regulated firm are equal to the fixed fee times the number of units sold. If a regulator continuously sets price equal to $AC + F$, profits are maximized by selling Q_r. If a smaller quantity Q' were sold, the pipeline would receive profits FQ' less than FQ_r. In order to sell more gas, the pipeline would have to offer a price less than $AC + F$, and would thus find its per-unit fee reduced. At the quantity Q_r, the price paid by the pipeline for the last unit of gas purchased, P^*, is greater than the price end-use consumers are willing to pay, P_r.

Representing regulation as a requirement to set prices at average cost plus a fixed markup is not unreasonable, but it does not tell the whole story. The purchased gas adjustment (PGA) mechanism allows the pipeline to make periodic increases in price, based on projections of increases in the cost of gas during some future time. These increases are ratified by the FERC without formal hearings. One purpose that the PGA serves is to avoid the need for frequent recourse to formal rate reviews in order to pass through cost increases over which the pipeline is presumed to have little control. This practice economizes on the cost of regulatory proceedings, provides sufficient time between reviews that "regulatory lag" can reduce the importance of A-J bias in investment decisions (Bailey and Coleman, 1971), and avoids the financial burden on pipelines caused by regulatory lag in a period of rising costs.

During the period between formal rate reviews, a pipeline with a PGA sells gas at prices that are approximately equal to its cost of purchased gas plus a fixed markup. The markup is adjusted at periodic rate reviews to reflect changes in the rate base, new throughput projections, and other costs not eligible for PGA treatment. If a pipeline succeeds in increasing its throughput during the time in which a particular price structure is in effect, it earns additional profits which it can in general retain under the practice of "no retroactive rulemaking." Thus the incentive is created to maximize throughput between rate reviews.

The size of this incentive depends on the time between formal rate reviews. If rate reviews are infrequent, the pipeline will behave as described by the model of a regulated firm selling gas at cost plus a fixed markup. If rate reviews are frequent, the markup itself will vary, as the pipeline is required to adjust its prices and input purchases to satisfy the regulatory constraint on its overall rate of return. This approximates the conditions assumed in the classic Averch-Johnson (A-J) theory. Under these conditions, no incentive is given to bid up the price of gas. Indeed, pipelines would always maintain a profitable wedge between the selling price of gas and their marginal cost. In short, in the A-J model the utility does not have an incentive to maximize output. Rather, as Wellisz (1963) has shown, under perfect regulation in which only a constraint on the rate of return that may be earned is continuously enforced, a utility's willingness to pay for gas is exactly its marginal revenue schedule in the market, less marginal transportation cost. Under such regulation, an incentive is created to overcapitalize—to expand capital investment beyond the point at which the cost of providing a given level of service is minimized. This overcapitalization tends to increase the pipeline's willingness to pay for gas, because it tends to increase the fixed cost and reduce the variable cost of transporting gas. On the other hand, any market power the pipeline has will reduce its willingness to pay for gas, because of the incentive to restrict output in order to obtain a higher price.

This argument can also be made more precise. The regulated pipeline faces a rising marginal cost of gas, $S'(G)$, and the price paid for prior units of gas is not affected by changes in purchases of gas. Invested capital is K, and the cost of capital is r. The allowed rate of return is $v > r$, and $Z(K,G)$ is sales of gas, which depends on capital and gas input G. Pipeline profits are $PZ(K,G) - rK - S(G)$ and are constrained by $PZ - S(G) < vK$. Under classic rate of return regulation the pipeline sets:

$$[P + P'Z]\, Z_G = S'(G)$$

Let $Z_G = 1$ for simplicity (the derivative of output with respect to gas input). The monopolist pipeline will therefore buy gas until the price of the last unit purchased is *less* than the selling price P; that is, it will supply less than the quantity consumers are willing to purchase at the marginal cost $S'(G)$. Thus rate-of-return regulation by itself creates no incentive to roll in high-cost gas.

Still another set of incentives would be created by regulatory lag if there were no PGA. In this case, the pipelines would face a fixed price between formal rate reviews, and would have an incentive to minimize cost. Paying more for gas than the regulated price at which it may be sold would reduce revenues until the price is adjusted. If rate reviews were infrequent, the losses involved between reviews could exceed the discounted value of future increases in revenues from larger sales, and gas would not be purchased at a cost greater than the price charged consumers.

These nuances of regulatory practice have been neglected in existing studies of the "bidding-up" problem. The size of the old gas cushion will not fully explain a pipeline's willingness to pay inflated prices for deregulated gas. The nature of PGA mechanisms and the length of time between rate reviews is also important, and the resulting bidding incentives can only be estimated after detailed examination of the regulatory process.

Market Equilibrium with Regulatory Lag

Some additional problems may be caused by the dynamics of price adjustment in the presence of long-term gas supply contracts and regulatory lag. They include erosion of the financial condition of pipelines, and a possibility of permanent market disequilibrium, with end-use markets failing to clear after field prices are deregulated. If a pipeline faces increasing average gas costs over time, as it will if demand is increasing or if deliveries of old, low-priced gas fall, regulatory lag will reduce the incentive to maximize sales. If the price at which gas may be sold is only readjusted at intervals to cover average purchased gas cost, the pipeline will incur losses equal to the increase in average cost above the legal price times the quantity of gas sold prior to rate adjustment. If rates were never adjusted, the pipeline would incur losses whenever it paid more for gas than the legal price at which it could be sold (less the average of other operating costs). But unless it did so, demand at the legal price would exceed supply, possibly in violation of contracts or obligation to serve. As with electric utilities, rising costs coupled with regulatory lag and an obligation to serve could generate a financially troubled gas transmission industry.

Combining simple, rigid contractual provisions with rigorous regulation of the profits of pipelines and distribution companies may also produce a state of chronic disequilibrium in downstream markets. An obligation to serve, created by law or by contract, may make it necessary for a pipeline to incur losses as long as it is physically possible to obtain supplies that would meet demand. If demand is growing and supplies of low-priced gas are dwindling, average costs may exceed the regulated retail price. If curtailments are not imposed, additional gas must be acquired through spot purchases or new contracts. If spot purchases can be made, average costs will rise and demand based on average cost will fall. If only contract purchases are possible, however, more gas will be contracted for to meet short-run demand than is needed in the long run when effects of higher prices on demand are seen. This is true even with PGAs, if a continuous obligation to serve is coupled with demand that is more elastic in the short run than in the long run. If long-term contracts must be signed to obtain gas and clear the market in the short run, take or pay provisions may create excess supply in the long run when demand responds more strongly to price. Temporary rents are required to cut off short-run hikes in demand, but they are not possible with PGAs and long-run contracts.

The existence of interruptible load suggests that the obligation to serve is not absolute. An increasing need for high-priced gas to meet high priority demand, coupled with regulatory lag, would lead a utility to shed industrial load rather than purchase gas at prices above resale prices. But this is nonprice rationing to curtail demand. Thus, tight constraints on pricing by regulated pipelines and long-term contracts may create either excess supply or nonprice rationing of gas despite field price deregulation.

Bias in Capital Investment

Pipelines and distribution companies can choose to meet seasonally fluctuating demand through investment in additional reserves, transmission capacity, or storage capacity. A-J theory predicts overcapitalization as a general consequence of rate-of-return regulation, but the choice among different types of capital investment has received little attention. Wellisz (1963) notes that a bias against efficiently high levels of investment in storage capacity will be created, though largely through the effect of regulation on output pricing. Aside from simple goldplating—building capital that is not expected to be used at all—it is difficult to envision other significant possibilities for bias in capital investment by pipelines. Callen's (1978) study suggests there is little flexibility to substitute rate base investment for other factor inputs in pipeline construction. Nevertheless, meeting irregular and uncertain demand is an essential feature of gas transmission. To evaluate fully whether an optimal degree of reliability is provided, possible biases in capital investment must be considered along with output pricing behavior—the subject explored in the next section. Then we turn to a discussion of reliability.

Price Discrimination

The literature on variable (for example, time of use) pricing, on interruptibility, and on price discrimination by a regulated monopolist can be easily applied to regulated, downstream gas markets. The literature has normative and positive strands, the former asking what should be done to maximize social surplus under various constraints, and the latter asking what a monopolist will do under various forms of regulation. Although a remarkably small fraction of the voluminous literature on the economics of regulation has examined natural gas pipelines and distribution companies, that which does concentrates on price discrimination issues.

Pricing of Natural Gas at Retail

One of the earliest articles developing a microeconomic model of firm behavior under regulatory constraint, by Wellisz (1963), treated the problem of pricing jurisdictional and nonjurisdictional gas sales by a pipeline. Wellisz examines two related questions.

The first is whether rate-of-return regulation leads to an efficient outcome. He shows that rate-of-return regulation creates an incentive for overinvestment in pipeline capacity, which is achieved by charging inefficiently low prices to peak users and inefficiently high prices to off-peak users.

In the simple model that Wellisz constructs, efficient pricing requires that if there is excess capacity in the off-peak period, the entire cost of constructing pipeline capacity should be paid by peak users. If the pipeline is fully utilized in the off-peak period, the price to off-peak users should include some of the cost of capacity, but the off-peak price should still be less than the peak (the peak demand curve is assumed to lie strictly above the off-peak demand curve).

The profit-maximizing monopolist, on the other hand, will charge both peak and off-peak users a price greater than the cost of purchasing gas and the variable cost of transportation, even if there would be excess capacity off-peak with efficient pricing. Rate-of-return regulation will cause the monopolist to reduce the price for peak customers, in order to increase peak demand and its justifiable investment in capacity, and will increase the price to off-peak customers. The total welfare loss will equal the deadweight loss of reduced off-peak demand and the cost of excess of capacity over what peak consumers would be willing to pay for peak gas supplies. Wellisz also notes that since rate-of-return regulation creates an incentive to expand the rate base, more capital-intensive transmission capacity will be chosen over storage options that would offer a savings on investment.

The second question Wellisz addresses is: What is the effect of setting rates for jurisdictional sales directly? He finds that the Atlantic Seaboard formula—which apportions fixed and variable costs among a pipeline's customers—

"limits economically justifiable nonregulated sales" and causes expensive pipeline capacity for meeting peak load to be substituted for less expensive storage capacity.

In fact, he shows that the Atlantic Seaboard formula creates incentives similar to those that a monopolist faces in setting off-peak prices, and leads to off-peak prices higher than opportunity cost. As a result, distribution companies, which face an artificially narrowed differential between peak and off-peak prices, are not encouraged to make optimal investments in storage.

Wellisz's analysis is interesting for a number of reasons. It illustrates the importance of realistic modelling of regulatory constraint, finding that the effects of regulating rates of return while allowing pipelines to set prices are different from the effects of regulating prices directly through a cost allocation formula. Moreover, the relation between pipelines and distribution companies is investigated explicitly. In doing so, Wellisz observes that final demand for gas can be distorted as it is translated into the derived demand of distribution companies, and that monopolistic pricing of off-peak gas by pipelines increases the peak-load resale price of distribution companies.

Rosenberg (1967) extends Wellisz's results by considering the problems arising from the Federal Power Commission (FPC) zone-allocation formula for determining capacity charges to consumers located at different points along a pipeline. In particular, the system-wide peak incorporated in the zone-allocation formula is found to shift a portion of downstream user cost to upstream customers. He demonstrates that, in general, if peak demand periods do not coincide at different points along a pipeline, efficient pricing may require that off-peak users at some location pay a portion, and possibly all, of the capacity costs.

Rosenberg's analysis amplifies the characterization of peak pricing first analyzed by Steiner (1957) and noted by Wellisz: When capacity serves demand in a number of periods, it is a public good from the point of view of use in different periods. Thus the classic Lindahl pricing formula applies. Prices must differ across periods, and must make demand equal to capacity in all periods (except those in which demand is less than capacity at a zero price). The sum of prices charged in each period must equal the marginal cost of additional capacity.

A monopolist may engage in price discrimination based on demand elasticities as well as distinguishing between peak and off-peak demand. MacAvoy and Noll (1973) concentrate not on peak pricing but on this form of price discrimination—contrasting prices in unregulated, industrial sales with those in regulated sales to retail utilities. They also explore how to formulate and assess the significance of "rivalry" among pipelines that might serve the same market.

The results that MacAvoy and Noll derive on the effects of regulation differ from those of Wellisz. They observe that throughout much of the 1960s and 1970s capacity has been greater than demand and that therefore capital investment has been fixed. This observation becomes the basis for their argument that, unless demand is inelastic, regulation will cause a pipeline to raise prices and lower sales in its jurisdictional market, compared with what they would be without regulation. Since capacity is fixed, the pipeline cannot expand its rate base to ease the rate-of-return constraint. Consequently, it must alter sales and prices in the regulated market to meet the profits constraint. If the pipeline were to lower prices, regulated sales would rise. But because pipeline capacity is fixed, unregulated sales would necessarily fall if the pipeline were initially operating at full capacity. Such action would also reduce profits in the unregulated market. Thus the pipeline's optimal choice is to raise prices and reduce sales in the regulated market until profits fall to the allowed amount, and to continue earning monopoly profits in the unregulated market.

This conclusion is correct given the assumptions, but it is actually a somewhat trivial result. It comes about simply because MacAvoy and Noll assume a very limited scope of choice for the pipeline. Since pipeline capital investment is assumed to be fixed, total profits allowed on regulated sales are fixed. Thus maximum feasible profits are equal to total allowed profits in the regulated market plus whatever the pipeline can earn in the unregulated market. Changing the price charged in the regulated market is assumed to have no effect on demand in the unregulated market, so that even if jurisdictional sales were deregulated, the pipeline would still be able to earn the same profits . *l* industrial sales. When more capacity becomes available for industrial sales because of reductions in retail sales, the pipeline is able to increase total profits.

Within these constraints, MacAvoy and Noll assume the pipeline is free to set prices as it wishes, even if retail utility sales are reduced enough to create excess capacity. This assumption implies, however, that regulators make no attempt to assess the fairness of rates by comparing direct sale prices with jurisdictional prices, and implies that cost allocation formulas, in which allowable prices on jurisdictional sales vary with the proportion of jurisdictional sales, are not used.

MacAvoy and Noll also discuss rivalry among pipelines. They hypothesize that rivalry has two effects: (1) the presence of an alternative source of supply limits a pipeline's ability to engage in a high degree of price discrimination and (2) the presence of rivalry increases the elasticity of demand facing any pipeline and thus reduces the price in that market. Specifically, they posit that when rivalry is present, pipelines will be unable to charge two-part tariffs designed to extract greater monopoly profits.[1] With less control of markets, they argue that "the elaborate structure of individual demand and commodity charges would likely reduce to single commodity prices" which would differ among customers only if marginal costs of transportation differ (MacAvoy and Noll, 1973, p. 215).

Even if rivalry is present, however, perfect competition is unlikely. A pipeline is likely to discriminate between easily recognizable categories of industrial and retail utility sales in setting prices. Prices will be set according to a formula similar to that of Bailey and White (1974). Relative prices in the two markets will satisfy:

$$P_1(1 + \frac{1}{e_1}) = P_2(1 + \frac{1}{e_2}) + [mct(q_1) - mct(q_2)]$$

where P_1 and P_2 are the prices charged in market 1 and market 2, respectively, and the last term is the difference in marginal transportation costs to the two markets. The elasticities of demand, e_1 and e_2, facing the pipeline will be larger than demand elasticities for the whole market if rivalry is present.

In their empirical analysis, MacAvoy and Noll find considerable differences in pricing between regulated and unregulated markets. Rivalry appears to constrain prices on industrial sales below the level predicted for a monopolist facing the market demand elasticity, but sales to retail utilities conform to the monopoly model of two-part tariffs. With two-part tariffs, the commodity charge is equal to marginal cost, independent of demand elasticities, and price discrimination is practiced by setting the demand charge. They point out that "(r)egulation, by providing the framework for two-part pricing, might be credited with reduced marginal prices at farther locations. There might have been as much rivalry on those transactions, in the absence of the Commission process, as on unregulated transactions—with the result that there would have been no two-part tariffs" (MacAvoy and Noll, 1973, p. 232). But, they conclude, the *average* price to retail utilities and pipelines' rates of return show no significant effects of regulation. MacAvoy and Noll's strong assumptions about the nature of the regulatory constraint, as noted above, should be kept in mind in evaluating these results.

General Results on Price Discrimination

The literature on the economics of price discrimination is not specialized to a particular industry, though authors seem to have electric or

1. By assuming that capacity exceeds peak demand, MacAvoy and Noll eliminate any other justification for two-part pricing as a means of allocating available capacity.

telephone utilities in mind. The natural gas industry has peculiar features that require care in applying general theoretical results, but considerable insight can be gained from the exercise.

Bailey and White (1974) provide a general statement of the theory of price discrimination that synthesizes the analysis of peak-load pricing, price discrimination by a monopolist, optimal departures from the marginal cost pricing, and multi-part pricing. They analyze peak prices designed to maximize welfare as well as prices designed to maximize a monopolist's profits.

WELFARE MAXIMIZATION. For efficient resource allocation, the peak price charged by a pipeline should exceed the off-peak price. Whether the off-peak charge should include any capacity charge depends on whether there is surplus off-peak capacity. If the pipeline's capacity is fully utilized in all periods, prices must be found that equate demand to capacity in each period. In the peak period, aggregate willingness to pay for gas is highest, and the highest price should be charged in that period. Whenever there is surplus capacity, no capacity charge should be included in the price (subject to the qualifications about pricing at different locations with noncoincident peaks noted by Rosenberg, 1967).[2]

Welfare maximization requires that there be no discrimination in pricing based on the relative elasticity of demand attributed to different customer classes, or times of use. (Note that the *level* of demand, not its *elasticity*, determines the optimal peak load, or more generally, Lindahl, price.)

If a pipeline faces increasing returns to scale (decreasing average costs with increasing output), prices whose sum equals marginal cost will result in negative profits. If two-part pricing is not utilized, prices must be set above marginal cost to allow the utility to break even. In the optimal departure from marginal cost pricing [a term coined by Baumol and Bradford (1970)], misallocation of resources involved in setting a break-even price is minimized. Misallocation of resources arises from reducing demand below the point at which willingness to pay for gas equals the marginal cost of service. Thus in a sense, the change in consumption of gas from the optimal level is a metric in which efficiency losses can be measured. As Blaydon, Magat and Thomas (1979) show, by increasing prices most for the class of customers whose demand will change least, the distortion in actual consumption of gas can be minimized. This leads to the inverse elasticity "rule": prices should be raised most for the class of customers with least elastic demand. Normally, it is thought that this type of price discrimination is made between customers of different types—residential, commercial, and industrial being the common categories. If the only discrimination allowed is between peak and off-peak use, the inverse elasticity rule suggests increasing the price during the period when demand is least elastic. If peak demand is more elastic than off-peak, the optimal departure from marginal cost pricing would reduce the peak price relative to the off-peak price. In the case of natural gas, however, it is likely that peak demand is less elastic than off-peak. Household gas use for heating is commonly held to be responsible for winter peaks in gas demand, while industrial demand for boiler fuel is constant through the year and a large component of off-peak use. Thus if home heating demand is inelastic and industrial demand very elastic, peak prices should be raised more than off-peak.

If two-part tariffs are used, marginal rates can be set at the optimal peak and off-peak tariffs, and an inframarginal price (which, by definition, does not affect demand) set such that the pipeline breaks even.

PRICING BY A MONOPOLIST. A monopolist capable of discriminating among customers will also follow an inverse elasticity rule of pricing. The monopolist, as Wellisz notes, will choose capacity on the basis of marginal revenue, rather than price. Peak capacity will be set at the point at which marginal revenue equals the marginal cost of additional capacity, and off-peak price will be set to maximize revenue. The less elastic the demand, the greater the difference between

2. The availability of storage for gas intended for peak use implies that capacity may be fully utilized in the off-peak period, and thus implies that off-peak rates should include a capacity charge (Gravelle, 1976).

marginal revenue and price. Thus if off-peak demand is perfectly elastic (as is alleged of industrial demand when the price of gas equals that of alternative fuels), the monopolist will increase the peak price above the welfare maximizing level, but leave the off-peak price undisturbed.

More generally, let e_p be the elasticity of peak demand and e_o the elasticity of off-peak, and assume that average variable cost is c and the average cost of capacity is rk. Then according to Bailey and White, the monopolist will charge a peak price

$$P_p = \frac{c + rk}{1 - \frac{1}{e_p}}$$

and an off-peak price

$$P_o = \frac{c}{1 - \frac{1}{e_o}}$$

If peak elasticity is equal to off-peak elasticity ($e_p = e_o$), the ratio of peak to off-peak prices will be the same as the ratio of marginal cost at peak to off-peak, and peak prices will always exceed off-peak prices. If off-peak demand is more elastic than peak ($e_p < e_o$), and marginal capacity costs are small relative to marginal operating expenses, optimal pricing will follow the inverse elasticity rule—since

$$\frac{dP}{de} = \frac{-1}{(e - 1)^2} < 0$$

The price will be lowest when e is largest in absolute value, that is, when demand is most elastic.[3]

The monopolist can always increase profits by discriminating directly among consumers, charging industrial customers a lower price than residential customers, and increasing only the residential price in a peak period. Multi-part pricing would make the monopolist better off yet, by allowing the monopolist to charge marginal cost prices c and $c + rk$ to all customers in peak and off-peak periods, and extracting the entire consumer's surplus through a customer charge (or any other sufficiently high inframarginal price). Since marginal cost pricing maximizes the sum of producer's and consumer's surplus, the adoption of efficient peak-load pricing maximizes the profits a monopolist could extract with perfect two-part pricing.

The unrestrained monopolist, if unable to discriminate except by time of use, is likely to charge a peak price greater than marginal cost of service and an off-peak price less, as long as peak demand is less elastic than off-peak. Rate-of-return regulation introduces the opposite bias. If peak and off-peak demand elasticities are the same, the monopolist under rate-of-return constraint will charge peak users less than the marginal cost of service. Thus, Bailey and White argue, contrary to Wellisz, that the off-peak price will not necessarily be increased by enlarging peak capacity. If this added capacity lowers the variable cost of providing off-peak service (lower fuel costs, say, because of more efficient compressors), the price to off-peak users may fall.

Reliability

The capacity provided by a regulated firm may be above the welfare maximizing level, depending on how much larger the rate base must be to absorb profits above those allowed.

Spence (1975b) provides an interesting variant on this argument when uncertainty of demand is introduced. If demand is uncertain, capacity may be insufficient to satisfy it at the predetermined peak price. As discussed more fully in the next section, either interruption of service or a price that varies *ex post* is required to equate demand with capacity. If interruption of service occurs, the monopolist will provide a degree of reliability, based on the *marginal* customer's willingness to pay. This is the correct criterion to apply for delivery, or quantity, of gas, but not the correct criterion for reliability.

3. For convenience, in exposition elasticity is defined as a positive number, by taking the absolute value (modulus) of the ratio of the percentage change in demand to percentage change in price. The larger the response of demand to a given change in price, the larger the elasticity.

Reliability (or more generally "quality" in Spence's model) should be based on *average* willingness to pay. Thus reliability will be undersupplied by the monopolist, and the bias introduced by rate-of-return regulation can serve to bring reliability up toward (or beyond) its optimal level.

The economic theory of agency—introduced in the previous chapter—demonstrates that an optimal degree of reliability will not be provided. Suppose a pipeline sells a block of gas (1 unit) to an interruptible customer at a price p, and that the wholesale price of gas plus transportation is g. Demand on the part of priority customers is a random variable, such that the probability that the pipeline will be able to deliver gas to the interruptible customer is s. The probability is $1 - s$ that the interruptible customer receives no gas. By increasing capacity, the pipeline can increase the probability that the interruptible customer will be served. The cost of achieving a service probability is $C(s)$. If the pipeline's objective is to maximize expected profits, it will choose a probability of service that satisfies:

$$C'(s) = p - g$$

Expected profits are

$$s(p - g) - C(s)$$

If gas is not available, the interruptible customer will pay $h > p$ for a substitute fuel. Thus the customer's expected costs are:

$$sp + (1 - s)h$$

The principal-agent problem then becomes one of choosing a price p that will minimize the customer's expected cost subject to providing the pipeline with some minimum profit V, given that s is chosen to satisfy $C'(s) = p - g$.

An overall optimum is obtained by maximizing

$$L = sp + (1 - s)h - u[s(p - g) - C(s) - V]$$

with respect to p and s (u is a Lagrangian multiplier). The first-order conditions are:

$$\frac{\partial L}{\partial s} = p - h - u(p - g - C') = 0$$

$$\frac{\partial L}{\partial p} = s - us = 0$$

$$s(p - g) - C(s) = V$$

An intuitive conclusion follows: $u = 1$, so that additions to pipeline benefits count equally with additions to customer benefits. The marginal cost of improving the probability of service $C'(s)$ should equal the cost of doing without service $h - g$ (the sum of pipeline and customer benefits forgone). And p should be set independently to divide profits between pipeline and customer.

In characterizing the agency problem, it is necessary to know what price induces the pipeline to set $C' = h - g$. Since the pipeline will satisfy $C' = p - g$, the price at which it sells gas must equal h; that is, all profits from gas use must go to the pipeline for it to choose an optimal degree of reliability.

This conclusion illustrates the point made by Spence: quality of service is determined by the demand of the marginal consumer. If there are two customers, with costs h and h' for alternative fuels, the optimal probability of service is given by $C'(s) = (h + h')/2$. But if both must be charged a uniform price, the pipeline will be able to serve both customers only if it charges a price equal to the smaller of h and h'. In that case it will provide too little capacity.

The pipeline could contract separately with each customer to supply gas at a price h or h' whenever gas is available, and pay the customer a fixed fee to induce each to hook up. This arrangement would satisfy the coinsurance-deductible theorem if the pipeline is risk neutral. A risk-averse pipeline should not bear all risk of capacity availability. A smaller share of the pipeline's contingent benefits, and a lower rebate to the customer, would cause the pipeline to choose too little capacity but would share risk more efficiently.

Integrating Approaches to Price Discrimination

Studies of the pricing behavior of the natural gas transmission industry differ most fundamentally in their treatment of price discrimination. Figure 5-3 depicts a convenient way to contrast these differences. Sales of gas can be divided into peak and off peak, as the columns are labelled. A pipeline may also be able to discriminate among sales to users with different elasticities of demand (the rows). It may also be able to practice a greater degree of price discrimination through two-part tariffs.

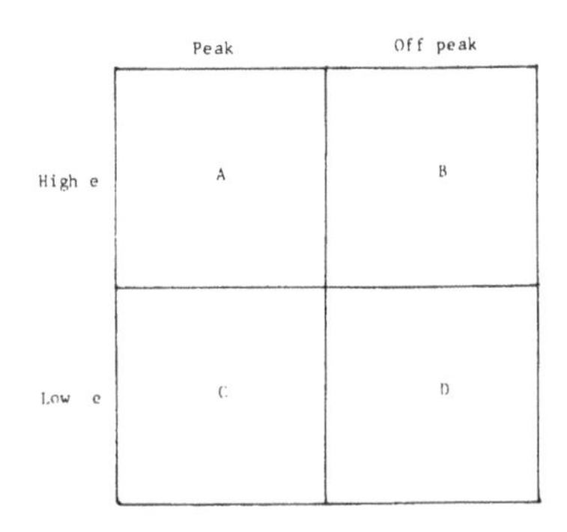

Figure 5-3. Price Discrimination Possibilities

The studies discussed above can be compared in terms of on which cells they focus their analysis of pricing behavior. Wellisz investigates "horizontal" price differences between peak and off-peak sales. Relative demand elasticities do not enter explicitly. MacAvoy and Noll (1973) focus on "vertical" price configurations between customer classes, not distinguishing peak from off-peak demand. Bailey and White (1974) make diagonal comparisons; they assume that peak and off-peak demands can have different elasticities.

To understand likely developments in direct industrial and retail gas sales, analyses of peak pricing and of price discrimination need to be further integrated. In reality, prices for gas utility services vary both by time of use *and* across customer class. Analysis of time of use pricing must also integrate the seasonal storage of gas. Price differentials between peak and off-peak use depend on the marginal cost of storage as well as the demand and capacity factors mentioned above. In order to analyze competition among pipelines, especially when it involves unregulated mainline industrial sales, it is particularly important to consider price discrimination across consumers. Even if seasonal variations in demand are ignored, problems arise in determining how to allocate the joint cost of pipeline capacity among users because rate-of-return regulation gives a clear incentive to allocate all fixed costs to the regulated sector.

This last point raises the issue of what economic theory suggests about the pricing behavior of a regulated firm that derives some of its earnings from operations not subject to regulatory constraint. Before turning to that issue, however, we explore more carefully pricing for reliability of service.

Variable Pricing with Stochastic Demand

Up to this point, the discussion has focused on input choices and optimal pricing when demand varies deterministically over time. But the presence of uncertainty in demand also affects the choices of a regulated firm and its pricing behavior. The critical new element introduced by stochastic demand is the problem of keeping future demand below available capacity. In this context, the issues of optimal reserve margin and storage, optimal interruption priorities, and variable "real-time" pricing need to be analyzed.

Time of use pricing has been studied intensively, primarily with regard to electricity, and may represent a unique example of successful advocacy by economists of improvement in regulation. More subtle schemes for real-time adjustments of prices to ration capacity in peak periods may be applicable to gas markets, but because of technological factors unique to gas, there are likely to be much greater costs of communication and metering (for example, signals cannot be sent along gas lines). Questions of optimal reserve margins may be more specialized to electricity because, in gas, the tradeoff between pipeline and production capacity, on one hand, and storage capacity on the other, is paramount. Some of these issues are addressed most appropriately in terms of the agency relation between pipelines and/or distribution companies and their customers.

It is useful to concentrate the analysis on mechanisms that help reveal preferences among service of different degrees of reliability. Because of the safety concerns involved in household gas use, interruption of that service is generally held to be impermissible. Fortunately, instantaneous peaks are not as dangerous to system reliability as they are in electricity. Rosenberg (1967) notes that even if stored gas is not available, line pack can be used to meet peaks for one to two days. These technical features of the gas transmission system allow interruptible industrial demand to buffer higher priority uses and for orderly curtailments to be made down a list of priority users. The question is how to construct that priority list, and how to reveal willingness to pay for reliability to distributors, pipelines, and producers. On this issue, two studies are of particular interest.

Panzar and Sibley (1978) developed a method to reveal demand for reliability and allocate service efficiently when the capacity constraint is binding. They argue that continuous rationing through price requires an ability to predict temperature-dependent willingness to pay (for electricity in their case, but the point applies equally to gas), and that sufficiently accurate estimation of that relation is impossible. The approach they suggest requires minimal information by allowing customers to "buy" capacity as well as service. It combines an optimal pricing scheme with a rudimentary form of load management. They propose that each consumer subscribe *ex ante* to a particular level of capacity and then be charged for the amount of subscribed capacity and for each unit actually consumed. (Supply to that user is physically limited to no more than the subscribed amount.) In essence, the consumer is envisioned as purchasing two complementary goods: the commodity itself and insurance against being rationed.

Panzar and Sibley show that in order to maximize welfare, the constraint on service, that is, subscribed capacity, should be priced at the marginal cost of capacity, and that consumption should be priced at marginal operating cost. Thus their two-part pricing scheme divides the peak price into its two components, charges for them separately, and allows customers to choose just how much service they will be allowed to consume when the system is forced into curtailment.

It should be noted that some of the simplicity of the Panzar–Sibley model comes from their assumption that the marginal cost of expanding capacity and the marginal cost of operation are constant. If, however, these vary with demand, the regulator has a complex problem finding a price schedule that will equate supply with demand. In practice, rates are designed by utilities and then approved or disapproved by commissions. Allowing the pipeline or utility to calculate marginal costs provides an opportunity to set discriminatory demand charges to extract monopoly rents. Panzar and Sibley do not address the question of what a monopolistic pipeline, regulated or unregulated, would charge. There is the additional problem, as suggested by MacAvoy and Noll, of whether a two-part tariff can be sustained when rivalry among pipelines exists.

The Panzar–Sibley model and proposal are not directly applicable to gas markets. Since the relevant peak for natural gas is a moving average of instantaneous consumption (perhaps over as much as three days), some sort of integrating device other than a peak-reading fuse would be required. An annual demand charge could be levied on households based on predicted use (size, Btu input, and so on), but an adverse selection problem would remain. Those who like warmth more than others would be more likely to pay the charge, thus making peak demand greater than predicted. It is also unclear whether industrial consumers with access to alternative fuels would simply switch off gas completely if required to pay for capacity to serve their maximum demand. Panzar and Sibley must also assume that every customer's maximum demand occurs at the same time. But many industrial gas customers voluntarily reduce demand to zero in peak periods to avoid demand charges. The interest of interruptible customers is in whether, and for how long, they will have to switch to alternative fuels.

Despite these qualifications, a useful idea for natural gas pricing emerges from the Panzar–Sibley approach. Separating prices for capacity from commodity charges can make it possible to determine what level of reliability customers are willing to pay for, and to ensure that when capacity is reached, gas is allocated to those most willing to pay. Yet there are still issues that need to be analyzed further: adverse selection, how to establish a market-determined price, and how pipelines can be motivated to develop a Panzar–Sibley type of rate structure.

As Panzar and Sibley suggest, extending their analysis to multiple periods could deal with the problem of noncoincident peak demands. Another approach would be to require payment for reliability: a demand charge, for example, that declined with the priority accorded a customer (or class of service). Such an approach would combine the insights of Panzar and Sibley with those of Tschirhart and Jen (1979).

In their analysis of how a monopolist will price interruptible service, Tschirhart and Jen propose that customers in any priority class pay one price for gas, whose magnitude depends on the probability of full service to that class, rather than paying two prices, one for a guaranteed supply up to some constraint and the other for gas actually consumed. The pricing rules they develop are not welfare-maximizing; exploring such an extension would be the next logical step. They also develop only a single price for each category of service, so that the optimality of a system such as theirs is questionable. Introducing two-part pricing, however, would not be difficult.

The Tschirhart and Jen formulation of the problem is of interest, rather than their ambiguous characterization of profit-maximizing prices. They assume the monopolist serves priority classes—with class 1 (households) interrupted last. Demand from class 1 is assumed to be stochastic, and demand from each other class is deterministic. Contracts are written with each class specifying a price of gas, a quantity to be supplied when the system is not operating at capacity, and a probability of interruption. Gas demand depends on the price of gas and on the quoted reliability (probability of interruption). Each class is assumed to contain consumers with identical price and reliability elasticities of demand. The monopolist chooses prices and reliabilities for each class and total capacity. A condition for market equilibrium is that actual reliability must be that which is quoted.

Profit maximization requires that no class obtain a degree of reliability higher than that quoted and that, for at least one class, price exceeds the sum of capacity and operating costs, while for all classes, price exceeds only operating cost. Capacity is adjusted until marginal cost equals marginal revenue. The monopolist will set prices in accord with the inverse elasticity rule, adjusted to ensure that quoted reliabilities are provided. If demand is independent of reliability, the lower the price any class (except class 1) faces, the lower will be its probability of service. (If the monopolist is to interrupt service and lose revenue, it will choose to interrupt the service that provides least revenues.)

When demand is affected by reliability of service, the monopolist has conflicting goals. "The optimum ordering (of interruptions) is one that tends towards low priorities for classes with price elastic demand (who face low prices) and high priorities for classes with reliability elastic demands" (Tschirhart and Jen, 1979, p. 258). But some customers may have higher price and reliability demands than others. In this case the lowest priority class may not be quoted the least reliable service.

It would be interesting to examine how this problem would be solved by a welfare-maximizing enterprise. It is likely that an even better scheme building on the insights of Panzar and Sibley would involve separate pricing for reliability and service.

Regulatory Boundaries and Cross-Subsidization

With the relaxation of curtailment rules and the increased potential for additional gas supplies brought about by upstream deregulation, interstate pipelines face new market opportunities. The drive for new markets is likely to be strongest in the industrial sector where direct sales are not subject to regulation. Thus unless the pipelines pushing into industrial markets shed their market positions in distribution company sales—which are regulated—the mix of regulated and unregulated activity in interstate gas transmission is likely to remain but shift toward greater unregulated activity. Of course, such a mix has long been the practice of a number of interstate pipelines, but if the shift in emphasis is a dramatic one, the so-called "regulatory boundary" problem could become more pronounced. One aspect of this problem closely associated with price discrimination and the other behavioral issues explored in this chapter is the potential for cross-subsidization.

Defining Cross-Subsidization: The Concept and What Causes It

The notion of cross-subsidization generally refers to the practice of pricing a good or service below its cost in one market while offsetting the loss incurred by inflating profits in another market. As such, cross-subsidization creates the two types of resource misallocation that are possible when price diverges from cost. With price below cost in the subsidized market, output is at a level where its cost exceeds its value in use; otherwise efficient competitors are deterred from entry into that market. In the other market, output is suboptimal; entry appears attractive to inefficient producers.

A temporary strategy of cross-subsidization may be pursued across unregulated markets to drive out weaker competitors or to inhibit entry. For such an action to be worthwhile in the long run, it must create sufficient market power for the predatory firm to set price above cost at some later time.

Cross-subsidization can also be practiced between unregulated and regulated markets. In contrast to transitory cross-subsidization that might occur across unregulated markets, long-run cross-subsidization can be profitable when it is practiced across regulatory boundaries. For example, an interstate pipeline can make direct sales at unregulated prices to industrial customers, and indirect sales at regulated prices to local distribution companies who serve residential and commercial customers. In its unregulated market, the pipeline faces competition that results in relatively elastic demand. In its regulated market, the pipeline is granted a quasi-franchise that can create monopoly power. There are several reasons why a pipeline in this situation may price below cost in its unregulated operations while pushing up its permitted profits in regulated activities. A precondition for such pricing is the existence of joint costs that regulators will have difficulty in allocating among regulated and unregulated activities. McKie's (1970) analysis of the problems posed by trying to separate a firm's regulated operations from its unregulated operations demonstrates that when the different types of risks and incentives which exist on either side of the regulatory boundary are coupled with unattributable costs, inefficiencies inevitably result.

The nature of the regulatory regime is also an important determinant of the prospects for cross-subsidization. As Noll and Rivlin (1973) note, the Averch and Johnson model of firm behavior under regulatory constraint is applicable in this context. If the regulated firm's rate of return is a function of its total capital investment, by cutting prices sufficiently low in its nonjurisdictional market and thus expanding sales, it can increase its capital base and thus its permitted total profits. The utility's losses in its unregulated operations can be more than offset by higher profits in its regulated market, which can be justified to the regulators with a larger rate base. (The underlying assumptions are that demand is more price elastic in the unregulated market, and that regulators cannot determine how much of the added capital investment is due to unregulated sales.)

Noll and Rivlin argue that utilities can facilitate cross-subsidization between regulatory boundaries by deliberately adopting technologies whose costs are difficult to attribute to jurisdictional versus nonjurisdictional operations. In short, there is less of an incentive to choose a least-cost technology than there is to choose a more costly one that permits a greater proportion of its costs to be attributed to regulated operations. When a greater proportion of total costs is unattributable, the utility's marginal cost in the unregulated market is lowered.

But there are other effects of rate-of-return regulation that might work against cross-subsidization. As we noted above, Wellisz's (1963) analysis of gas pipelines suggests that in order to increase its rate base, and thereby its profits, the regulated firm will set prices for regulated sales relatively lower than for unregulated sales. He reasons that setting lower prices for peak customers—generally residential and commercial accounts supplied through regulated sales to local distribution companies—will create a need for extra peak load capacity. When peak customers are also those served in a regulated market, the rate base can be expanded by reducing prices for regulated sales relative to unregulated sales. Whether there will be cross-subsidization, and in what direction, depends on which incentive dominates. Interestingly, MacAvoy and Noll (1973) find that, after controlling for cost (and demand) differences, prices in regulated sales by interstate gas pipelines were not significantly lower than prices in unregulated sales.

Nonetheless, the analyses of Wellisz and MacAvoy and Noll are important in the context of cross-subsidization. On the one hand, they force a reminder that cross-subsidization cannot be described in terms of relative prices without regard to relative costs; on the other, that there is a link between rate structure and cross-subsidization—not only from the perspective of price discrimination among customer classes, but in terms of how the absolute price level for any particular class relates to the composition of fixed and variable costs attributed to that class.

This last point suggests that beyond the existence per se of regulatory boundaries and beyond the nature of the regulatory regime itself, cross-subsidization across jurisdictional and nonjurisdictional markets may derive from the way institutional rules regarding the allocation of costs are structured. Garfield and Lovejoy (1969) argue, for example, that the Atlantic Seaboard formula, by requiring an insufficient allocation of common fixed costs—which are largely a function of peak load capacity—to residential and commercial customers, in effect provided a subsidy to these users. Note, however, that here it is the regulated market that is being subsidized, rather than unregulated customers, which would be the effect of the A-J model alone.

The incremental pricing provision of NGPA is another institutional determinant of relative prices in regulated and unregulated sales and thus indirectly related to the prospects for cross-subsidization. Again, the effect runs the other way than that identified with the A-J paradigm. Under the incremental pricing scheme, large industrial customers are required to pay a disproportionate share of the total costs of gas rather than a simple weighted average that was inherent in "rolled-in pricing."

Identifying Cross-Subsidization

In unregulated manufacturing businesses, cross-subsidization may be practiced, in theory at least, by a (conglomerate) firm that produces two very different products, say farm equipment and cardboard packaging. To offset losses in its farm equipment division brought about by pricing such equipment below cost in order to drive out rivals, the firm may be able to earn "above-normal" profits in cardboard packaging—provided the latter's market structure permits (that is, if the packaging market is concentrated and has high barriers to entry). Cross-subsidization also could be practical, in theory, across two geographic markets in which the same good is sold. For example, a bread manufacturer could offset losses in its New York operations with excess profits earned in its Boston market.

In the first example, identifying the extent of cross-subsidization will be relatively simple because the products are produced and sold separately. Knowledge of the separate cost functions and the demand conditions for each product makes it possible to calculate price-cost margins. In the second example, there is a better chance that there will be some costs which are difficult to attribute to one market or the other. Identifying cross-subsidization in this case will be correspondingly more difficult.

The products of utilities tend to be services and will vary along a number of dimensions, including reliability. Thus, as discussed above, some gas customers, for example, large industrial users, may purchase gas on an "interruptible" basis, whereas others, residential consumers, for example, tend to buy gas on a "firm" basis. But what makes identifying cross-subsidization by utilities difficult is not so much these subtle product differences as the fact that common costs tend to be a high proportion of total costs. Different prices for utility services do not imply cross-subsidization; nor do equal prices, that is, "postage stamp" rates, imply the absence of cross-subsidization. In short, because the appropriate yardstick for measuring cross-subsidization is relative price-cost margins rather than relative prices themselves, and because the allocation of common costs of production must be made on some arbitrary basis, identification of cross-subsidization by firms in the utility sector is made that much more difficult.

Theory suggests two possible standards for measuring cross-subsidization where there are joint costs of production: the "marginal" standard and the "stand alone" standard. By the marginal standard, if the revenue generated by increased sales of a product covers the cost of producing that additional product, then the market in question is subsidy-free. But what should be used as a measure of the cost of additional production? Baumol and Walton (1973) argue that short-run incremental cost is the proper measure. Noll and Rivlin (1973), however, contend that "fully distributed" cost—incremental cost plus some share of common production cost—is more appropriate since it reflects long-run cost conditions which are more relevant in potential competitors' decisions about entry. This debate about short-run versus long-run cost has extended into the general discussion of how to identify predatory pricing [see the exchange between Areeda and Turner (1975) and Scherer (1976)].

The stand-alone test compares the cost of joint production with the cost of producing each product/service independently. Zajac (1972), for example, notes that a long-standing presumption in analysis of equity pricing is that "no customer group should pay higher prices than it would pay by itself." If there are increasing returns to scale to joint production—as is likely to be the case in gas transmission—the stand-alone test will generally result in higher costs than the fully distributed criterion, even if all common costs are shifted to one market. Thus, in most circumstances, if the stand-alone test is adopted, there would seem to be a bias toward the finding that subsidization exists.

There is no consensus in the literature as to which cost standard is superior. While the stand-alone principle may be intuitively appealing, in terms of economic reasoning it is inferior to either the incremental or fully distributed cost criterion; but the debate over the latter two does not appear to be converging. In any event, there would still be no development of a standard for assessing cross-subsidization even if one of these criteria emerged as a winner: Identification of costs is only half of the cross-subsidization issue; prices are equally important. Yet as demonstrated throughout this chapter, analysis of utility price setting is complex enough when the focus is on the subtleties involved in distinguishing between the determinants of discriminatory and peak/off-peak pricing behavior. The analysis becomes much more difficult when cross-subsidization pricing is included.

Overall, this brief discussion suggests that there are numerous offsetting influences to the practice of cross-subsidization. To be sure, there are incentives for a utility operating across regulatory boundaries to assign a disproportionate share of common costs to its regulated operations in order to inflate its earnings. Whether or not this actually happens is probably largely a function of the firm's ability to fool the regulatory authorities.

One way that regulators might try to prevent cross-subsidization is to focus on the rate of return after lumping regulated and nonregulated revenues together. But as McKie points out, this might have the same effect as extending regulation beyond its appropriate boundary and thus creating additional inefficiencies.

Studies of other regulated industries, such as telecommunications, suggest that there are incentives for cross-subsidization between regulated and unregulated markets. But the central lesson from the theoretical literature and its applications to natural gas transmission is that these incentives are subtle and possibly offset by other elements of pricing behavior.

An Overview of Lessons from the Literature on Regulation

The economic theory of regulation provides considerable insight into how continuing regulation (both federal and state) might affect the behavior of pipelines and distribution companies in the wake of field price deregulation. It suggests that the behavior of regulated firms with respect to input and output choices and price setting is sensitive to changes in the overall economic environment, in general, and in the form of the regulatory constraint in particular.

The general theme which emerges is that there are countervailing influences which will determine how well downstream participants translate the value of gas to end users into derived demand for gas in the field. Field price deregulation will tend to raise prices to all final consumers, and to make gas available to formerly curtailed users. But the pattern of prices at the burner tip could move toward or away from the optimal configuration depending on how pipelines, distribution companies, and regulators interact. Distortions or inefficiencies seem inevitable because regulatory authorities cannot control all the actions of regulated firms; but even if they could, this would impose such rigid constraints on managerial discretion that incentives for efficient utility operation would likely disappear.

Yet, identifying the sources of potential inefficiencies is complex because of the nature of the interaction between the participants: Decisions by regulators are affected by actions of the regulated firm, and firm behavior is influenced by its ability to affect the regulatory process. This interaction is thus a crucial variable determining how downstream behavior will be affected by upstream regulation. It is also a variable that must be specified in analyzing potential improvements in downstream regulation. Simply comparing predictions of outcomes under regulation to "optimal" outcomes can serve to identify potential problems, but cannot alone identify what changes should be made. Unless an unregulated market can be expected to satisfy all the conditions for optimality, some compromise among administrative costs of regulation and various distortionary effects is needed.

With or without field price deregulation, downstream gas markets will differ from the competitive ideal. They are characterized by significant quasi-fixed costs, significant concentration of buyers or sellers in some markets, and transaction costs that make complete risk-spreading and matching of supply to demand under all contingencies impossible. Regulation of the markets overlays these market imperfections, correcting some and exacerbating others.

The market imperfections and regulatory constraints in downstream markets will affect how pipelines bid for additional gas supplies, but careful scrutiny of all the forces that determine how a regulated firm makes input choices suggests that deregulated prices cannot be sustained above a market-clearing level by the bidding process alone. Pipelines and distribution companies will still exhibit discriminatory rate structures as field prices rise, although there may be slight changes. The prospects for cross-subsidization by pipelines operating in both regulated and unregulated markets could become greater with upstream deregulation, but realistic modelling of regulated firm behavior suggests that the problem may not be pronounced.

As in the previous chapter, where analysis of contracting and principal-agent problems reveals completeness of natural gas markets as a crucial question, so too does it emerge in the context of behavioral problems. Providing an optimal degree of reliability seems to depend on whether gas consumers can purchase reliability separately from delivery. And, for efficient time-of-use pricing, separate transactions must be made for peak and off-peak sales. In this sense, analysis of the effect of upstream deregulation on downstream behavior parallels that of transactional and structural questions.

Research Design

There is a wealth of possibilities for applying the analytical techniques and results described in this chapter to natural gas markets. Both theory and data must be developed before it will be possible to predict with confidence how pipelines and distribution companies will behave under the constraints of continued downstream regulation.

First in importance is incorporating Joskow's approach to analysis of regulation into theoretical models of regulated pipelines and distributors. To formulate such models it is necessary to make realistic assumptions about the nature of the regulatory process. Empirical study of current practices of the FERC and state agencies can provide some insight, but changes in regulatory practice in response to the changing conditions created by field price deregulation are likely. As in the case of the electric utility industry described by Joskow (1974), regulators are likely to look for more efficient processes if there is a serious conflict between old ways of regulating and new market conditions.

This possibility introduces an interesting simultaneity in the research design, because the response of regulated firms to different modes of regulation is likely to be an important factor in the choice among modes. Thus it is desirable to analyze the implications of alternative hypothetical processes of regulation in addition to current processes. By doing so, the analysis can provide normative results useful to policy makers concerned about how best to regulate the future natural gas market.

Such a theoretical model should treat not only the standard questions of bias in input choices, but also the risk-bearing propensities and the preferences among end-use pricing methods that regulation induces in regulated firms.

Normative models of time-of-service pricing and of cost allocation among customer classes based on the technical characteristic of gas markets are also required. These can be developed along the lines of studies of optimal pricing of electricity. Reliability of service is a particularly neglected subject. With the end of forced curtailments, pipelines and distributors face a choice of how much to invest in reserves, transmission, and storage capacity to increase the reliability of service to industrial customers. The literature suggests that demand may be responsive to service reliability. Research on the demand for reliable service and on devising regulatory practices that match willingness to pay and costs of reliability thus would be especially valuable.

Finally, the issue of pricing by a firm that serves regulated and unregulated markets is a critical one in the gas industry, in which interstate pipelines have a choice between unregulated industrial sales and regulated sales to local distribution companies. Such research should investigate theoretical criteria for and practical means of ascertaining whether industrial sales are made at subsidy-free prices.

6

Changes in Downstream Regulatory Regimes

Introduction

Each of the three previous chapters described analytical methods appropriate for studying some feature of natural gas markets. The purpose of such studies would be to obtain a sufficient understanding of the forces at work in natural gas markets to predict their evolution, to assess their performance, and to design remedial policies (if needed). In this chapter we describe two issues that are both broad and likely to remain unresolved for some time, so as to illustrate how the products of a systematic research effort could be used. The issues addressed relate to: (1) the regulatory and carrier status of pipelines, and (2) rules regarding the allocation of costs in downstream rate structure.

Natural gas pipelines operate predominantly as "private carriers," not only transporting gas from field markets to the city gate, but also acting as buyers and resellers of the gas they transport. They have done so in markets in which the level of field prices, for all practical purposes, has been set by government regulation. With the freer upstream markets that will be created by NGPA, greater field price uncertainty will increase the risks to private carriers. Pipelines are likely to prefer to operate increasingly as "contract carriers," no longer taking ownership of the gas they transport. Such a shift will serve to reduce a pipeline's risk of being unable to sell gas at a price above its acquisition cost. In addition, the greater uncertainty brought about by upstream deregulation may create incentives for gas producers and distributors to seek "common carriage" regulation of pipelines to ensure better pipeline access. However, common carriage could leave important economic functions now performed by pipelines without a home. In order to evaluate alternatives to private carrier status, it is necessary to examine how different regulatory/carrier statuses change risks and incentives not only for pipelines but also for upstream and downstream firms.

Another regulatory practice that may change is in the setting of various formulas to apportion common fixed costs among customer classes in the context of a two-part tariff structure. The trend has been to allocate an increasing portion of fixed costs to the commodity charge and to reduce the rate differential between high load-factor and low load-factor customers. NGPA's incremental pricing provisions provide an additional impetus to changes in cost allocation downstream. These provisions result in pricing formulas under which large, primarily industrial consumers would be required to pay a greater proportion of the acquisition costs of expensive "new" gas than do residential customers. But a "cap" on industrial prices under the incremental pricing provisions creates an incentive for state regulatory commissions to increase the allocation of fixed costs to incrementally priced customers so as to avoid allocation of "incremental" costs to any customers in their jurisdiction. The literature on two-part tariff design reviewed in the preceding chapter suggests that an evaluation of changes in regulatory practice requires an analytical framework that permits a rigorous assessment of effects of changes in cost allocation on economic welfare.

Changes in the Regulatory/Carrier Status of Pipelines

Most natural gas pipelines operate as "private carriers."[1] In this role, they not only transport gas from the field to the city gate, but they actually own the gas they transport; that is, they buy gas at the wellhead and resell it to distributors. Alternatively, some gas pipelines have operated in part as "contract carriers," where private parties—producers (upstream from the pipelines) and distributors (downstream from the pipelines)—contract for a pipeline's transportation services. Under these circumstances, pipelines may carry out a brokerage function, matching up buyers with sellers, but do not take on a merchant, buyer/reseller role. They transport gas that they do not own. Regardless which of these carrier modes a gas pipeline has chosen—and a given pipeline will practice both—the pipeline's rate of return has been subject to government regulation; pipelines that transport gas across state lines are regulated at the federal level and intrastate pipelines are subject to state regulation. But, government regulation is neither a necessary nor sufficient condition for a pipeline to choose to be a private carrier or contract carrier. While regulation is an exogenous institutional constraint, choice of carrier status is endogenously determined by both economic and institutional considerations. An unregulated pipeline could choose to operate as a private or contract carrier.

By contrast, there are two types of carrier status that involve government regulation. "Common carriage" mandates, by state, federal, or local law, that a carrier must transport commodities—often subject to quality and/or volumetric specifications—owned by any firm willing to pay the established, regulated rate. Railways, oil pipelines, and some forms of motor transport are subject to this status. In the natural gas context, common carrier status could require, by law, that a pipeline transport gas for any producer or distributor at a regulated tariff.

While private or contract carriers have some freedom to choose between those two modes, once common carrier status is conferred—usually by legislative (congressional) action—there is little flexibility to change the mode of operation. In some industries, however, common carriers do in fact act like private or contract carriers when they transport their own commodities. An example is an oil pipeline that is part of an integrated company involved in either crude oil production or retail distribution of refined products.

Most fixed-route transport systems in the United States have common carrier status established by the Interstate Commerce and Hepburn Acts. Three rationales are usually cited. First, in industries where production activity and end-use markets are great distances from each other (for example, agriculture and natural resource extraction), common carriage is seen as a way of ensuring equal access to transport facilities at rates reflecting transport costs, thus overcoming potentially discriminatory behavior by transporters which might favor some producers or distributors at the expense of others. Second, common carriage regulation prevents "ruinous competition" between carriers in industries with large fixed costs (for example, railroads). And finally, to the extent that there are large economies of scale, that is, a condition of declining

1. Additional discussion of changes in regulatory/carrier status is found in Broadman, Montgomery and Russell (1982).

unit costs over the feasible range of output, the relevant market can support only a few carriers, giving rise to tight oligopolistic or possibly monopolistic pricing. However, while there may be a strong rationale for regulating decreasing cost industries, such regulation can be obtained in a carriage status other than common carriage.

Each type of carrier status confers certain risks and incentives on the carrier and in turn on parties upstream and downstream. For example, private carriers of storable goods will directly bear the risks associated with inventory management. A decline in the value of the commodity and poor inventory decisions will necessarily affect a private carrier's balance sheet. Contract carriers, on the other hand, do not take on a merchant function and can charge storage fees and share inventory risk with upstream and downstream parties. To take another example, private carriers have every incentive to perform a brokerage function to minimize the risk of buying a commodity that cannot be sold. Common carriers have no incentive, while contract carriers generally have some lesser interest in matching buyers and sellers, and may yield the brokerage role to a separate party. One incentive that is generally shared by all carrier modes, however, is to arrange for long-term utilization of their capital stock. For instance, oil and gas pipelines—due to both institutional incentives (that is, service certification requirements) and economic incentives (that is, to maintain steady pressure and to spread fixed costs)—are likely to seek out long-run transactional arrangements either through contracts or through vertical integration. In oil and gas pipelines, this incentive may be particularly strong because, unlike other transport industries (for example, railroad and motor transport), these systems must bear the risk associated with being able to transport only one very specific commodity along a fixed route.

In general, parties upstream and downstream from carriers face different types of risks. Producers located away from markets bear risks in getting their products to points of final consumption. Thus, they have an incentive to secure adequate capacity on a line at favorable rates—either a lower tariff on a contract carrier or a high price paid for gas by private carrier. End users may bear the risk of facing monopoly power, which may be greater when they face a private carrier.

Field price decontrol, by altering concern from supply to price, will likely change the risks and incentives associated with various types of carrier status for pipelines. Prior to NGPA, private carriers—the rule rather than the exception in the gas transmission industry—faced relatively low risks in purchasing and reselling. The purchase price was relatively fixed and most of their gas was sold to rate-of-return regulated utilities who likely did not have strong incentives to bargain down the price of their gas inputs because of the nature of the regulation (see chapter 5) and fuel price adjustment clauses. Under wellhead deregulation, however, pipelines are more exposed to fluctuations in field prices. Moreover, the end-use demand they face is more sensitive to changes in prices of alternative fuels as the price of gas rises. From the pipeline perspective, private carriage is likely to be seen as less advantageous. Contract carriage can reduce exposure to risk.

More generally, the uncertainty endemic to a freer market will emerge for the first time in the contemporary gas industry. Pipelines as well as producers and distributors in varying degrees will face greater uncertainty. As a result, some producers and distributors have an incentive to move away from private toward contract carriage.

Yet field price deregulation is also likely to create an incentive for some producers and distributors to favor a change in pipeline regulation such as subjecting pipelines to common carrier status. From a small producer's perspective, a shift to contract carriage helps reduce exposure to pipeline monopsony power, but it does not guarantee access to a line. Common carriage would—at least theoretically—shield smaller producers from pipeline discriminatory practices. At the other end, some distributors are fearful of monopoly pricing. Contract carriage may reduce the potential for pipelines to engage in this type of conduct. Under this status, the pipelines do not own or sell gas, and "monopoly pricing" could only arise in the form of unduly high charges for transportation services. In contrast, common carriage presumably would provide greater assurance that transportation charges would be "fair," in the sense of being uniform within classes of transaction established by the regulators. In the final analysis, however, producers and distributors face a tradeoff on the contract-common carriage issue between the risks associated with not having pipelines as brokers and the risks associated with not having legal guarantees against monopsony/monopoly behavior. Though pipelines would likely continue to carry out the brokerage function under contract carrier status, producers and distributors would be forced to take on that role themselves under a regime of common carriage unless independent brokerage enterprises emerge.

In addition, if contractual arrangements give way to vertical integration (see chapter 4), independent producers, and to some extent distributors, may be exposed to even greater risks of discriminatory behavior on the part of pipelines who also engage in oil production. Controlling access to transportation could enable these pipelines to create barriers to entry into particular producing regions. As the economic history of the oil industry suggests, in this case the pressure by producers and distributors for common carrier regulation of gas transportation could become strong. It is important to point out, as we do in chapter 4, that there is a simultaneity problem underlying this analysis. Just as the regulatory/carrier status of pipelines will be influenced by changes in the structure of transactional arrangements after field price deregulation, the incentives for vertical integration will depend significantly on the regulatory/carrier status of pipelines. For example, private carriers will generally have a stronger incentive for integration than contract or common carriers.

It should also be noted, however, that problems in long-term contracting between producers and pipelines would not be eliminated by a move away from private carriage. If producers and distributors were to enter into similar long-term contracts, market rigidities could still remain and price signals from consumers to producers could remain distorted.

Another indirect effect of upstream deregulation is related to that mentioned immediately above, but is perhaps more political than economic. The higher burner tip prices brought about by NGPA may provoke a consumer/congressional response. Pipelines could be made a scapegoat and common carrier regulation seen as the most politically acceptable remedy. On the other hand, some have argued (see Tussing and Barlow, 1982) that with field price deregulation allowing competition at the wellhead and interfuel competition at the burner tip, pipelines could without harm be freed from regulation entirely. Complete deregulation of pipelines would obviously rule out common carriage status.

Field price deregulation could well trigger further consideration of pipeline deregulation by removing the need to prevent pipelines from extracting rents created by chronic gas shortages. An evaluation of this alternative requires answers to the questions of pipeline market power raised in chapter 3.

Changes in Rules for Cost Allocation in Pipeline Rate Design

Pipeline regulation not only affects the level of prices by limiting the rates of return on invested capital, but also their structure. In general, regulations require that the revenues collected from a given customer class be proportional to that class's share of the cost of production, that is (at least), cover the cost of providing service to that class. The two types of costs that must be allocated among a pipeline's customers are fixed costs (the costs of pipes, compressors, and other equipment, as well as taxes and interest payments on debt) and variable costs (the pipeline's costs of acquiring gas inputs in the field and operation and maintenance costs). Calculating a given customer class's "fair share" of total costs thus necessitates the apportionment of (fixed) costs that are common to the various customer classes—an accounting problem for which no clear-cut economic criteria provide a solution. A two-part tariff is a means for classifying the production costs (both fixed and variable) into two components, a "demand charge," reflecting volumes of gas used at peak periods, and a "commodity charge," reflecting annual average volumes of gas consumption.

Various customer classes, for example those with a low ratio of peak to average use versus those with a high ratio, benefit differently from alternative distributions of costs within a pipeline's rate structure. As a result, the rules for cost allocation have been subject to considerable controversy and to modification over the past forty years. The evolution of the rules regarding gas pipeline ratemaking can be perhaps best traced using a simple equation. A pipeline's rate design is subject to the constraint that the revenue from each market cover fixed and variable costs such that:

$$\sum_{i=1}^{n} \frac{(1-a)F}{Q} + V_i q_i + aFp_i = C$$

where

q_i = total sales in the ith market

$$Q = \sum_{i=1}^{n} q_i = \text{total pipeline sales}$$

F = total fixed costs
V_i = variable costs attributable to the ith market
p_i = proportion of peak sales in the ith market such that

$$\sum_{i=1}^{n} p_i = 1$$

C = total costs
a = the parameter ($0 < a < 1$) by which fixed costs are allocated under rules set by the authorities.

The first component on the left-hand side of the equation is the commodity charge and varies directly with the level of gas consumption, q_i. The second component is the demand charge and is a function of the ith market's consumption during peak periods. When the regulatory parameter is set at zero, all costs (both fixed and variable) are assigned to the commodity charge; this is known as the "volumetric formula" because there is no demand charge in this case. If $a = 1$, the "fixed-variable formula," 100 percent of fixed costs is assigned to the demand charge and the commodity charge is comprised only of variable costs. In general, high load-factor customers (those with a low ratio of peak to average use) would prefer a greater portion of fixed costs in the demand charge and a smaller portion in the commodity charge. The result, a higher value of a, would reduce their average cost of service. Low load-factor customers would achieve lower unit costs if a smaller portion of fixed costs would be allocated to the demand charge, and consequently they prefer a lower value of a. Moreover, the choice of a cost allocation scheme influences not only the apportionment of total system costs among jurisdictional customer classes, but also the apportionment among jurisdictional and nonjurisdictional consumers. Pipelines would generally prefer a higher percentage of fixed costs allocated to the demand charge (a higher a) because it would tend to reduce the costs apportioned to high volume industrial sales. Such sales, to the extent they are direct wholesale transactions, are unregulated. As a result they can be made at prices above the allocated costs. In regulated sales, however, prices can be increased only if costs allocated to those sales are increased.

In 1952 the Federal Power Commission adopted the so-called Atlantic Seaboard formula wherein 50 percent of a pipeline's fixed costs were to be allocated to the commodity charge and 50 percent to the demand charge, that is, $a = 0.5$. Low load-factor customers faced higher average costs than those with high load factors. The effect of the Seaboard formula was to set the commodity charge above marginal cost and the demand charge below marginal cost. Distributors' rate structures tended to reflect this cost allocation. Compared with an ideal peak load price system in which all fixed costs appeared in the demand charge ($a = 1$), residential and commercial (peak) end users paid lower rates and industrial (off-peak) consumers higher rates. This rate design thus tended to a price structure inconsistent with that suggested by the literature on peak-load pricing. Since greater peak use causes greater capacity to be installed, peak users ought to cover most of the cost of capacity. Hence, peak users should bear a high proportion of capacity costs.

Modifications to the formula that place a greater proportion of fixed costs into the demand component were at times allowed to ward off competition with coal in the relatively elastic industrial market. Those pipelines facing the greatest interfuel competition received the largest variance from the rule. In this case, adherence to the principles of "value-of-service" pricing (that is, pricing according to demand elasticity), also served to move the formula closer to true peak load pricing. Murry's (1973) analysis, building on Wellisz (1963), suggests that the interaction between the rate-of-return constraint embodied in utility-type pipeline regulation and the Seaboard formula may also be important. Because the Seaboard formula allocates some fixed costs to the volumetrically determined commodity charge, the price for off-peak (mainly industrial) users is greater than opportunity costs. But insofar as regulation constrains the pipeline to a fixed rate of return, the higher profits earned off-peak by setting average revenue above opportunity costs must be offset by pricing peak sales below opportunity cost.

In 1973, twenty years after the adoption and partial erosion of the Seaboard ruling, the FPC adopted the United formula, which allocates 75 percent of fixed costs to the commodity charge, that is, $a = 0.25$. This reversed the trend toward reducing the allocation of fixed costs to the commodity charge. The difference in average rates between high load-factor and low load-factor customers and between industrial and residential end users embodied, at least nominally, in the Seaboard formula, has thus been diminished. In short, United runs contrary to the principles of efficient pricing. The FPC's rationale for this cost allocation scheme was to reduce industrial demand during a period of gas shortages in order to free up supplies for residential and commercial consumers. Although the United decision withstood a major court challenge in 1975, it is presently in a state of legal uncertainty (Pierce, 1980). Still, it has significant implications for the ultimate pricing of natural gas and the effects of price deregulation under NGPA.

The effects of the United formula are complicated by the impact of NGPA on the computation of variable costs. Until NGPA, the allocation of variable costs among a pipeline's customers was relatively straightforward, because prices were based uniformly on the average cost of gas inputs. Yet as differences in the cost of gas became more pronounced, so too did the inefficiency of rolled-in pricing. As Camm (1978), among others, has pointed out, the rolling-in of more expensive gas with low-cost gas leads to an average price that is below marginal gas acquisition or replacement costs and results in resource misallocation. Moreover, the marginal unit of gas is purchased by consumers at a price below its value in use.

NGPA adopts a regime of "incremental" pricing and institutes a major modification in the way variable costs are incorporated in pipeline ratemaking. It forces interstate pipelines and local distribution companies to allocate the expensive "new" gas to "low priority" large (greater than 300 mcf/day) customers (primarily the industrial market) and of low-cost "old" gas to "high priority" (residential) consumers. Thus consumers are partitioned into two classes, a high tier group that is required to pay a greater proportion of gas acquisition costs and a low tier group that is required to pay a disproportionately smaller fraction of the costs of acquiring gas. However, the provisions stipulate that as soon as a high tier customer faces a gas price that is equivalent to that of fuel oil, the pipeline and, in turn, the distributor serving that customer are no longer required by the FERC to subject it to an incrementally priced rate structure. As Pierce notes, at that point NGPA permits FERC some discretion as to whether that customer's rate structure can revert back to one consonant with rolled-in pricing or to "more radical" forms of incremental pricing.

It is through its incremental pricing provisions, which directly affect variable cost allocation in pipeline rate design, that the NGPA is likely to influence indirectly the allocation of fixed costs within the constraints imposed by United. However, identification of the possible NGPA-induced changes in the *rules* governing the allocation of costs is much more difficult. The regulatory authorities—both at the federal and state levels—are not required by NGPA to modify cost allocation rules. And distributors and the state public utility commissions are prohibited from instituting any changes that might neutralize the increase in rates that the incremental pricing provisions bring about—for example, by decreasing the allocation of fixed costs to customers subject to incremental pricing. But as Pierce argues, there are not likely to be any such offsetting changes through a decrease in fixed cost allocation. Indeed, to the extent that a distributor increases the allocation of fixed costs to its customers subject to incremental pricing, it is able to minimize the total cost of its gas inputs. This reduction in total costs will come about because the pipeline from whom the distributor purchases gas can, in turn, increase its allocation of fixed costs in the distributor's rates and thus decrease its own incremental pricing charges—charges that are the distributor's costs. The incentive for the distributor to increase the allocation of fixed costs to its incrementally priced customers will be in force, of course, only until the incremental price reaches the alternative fuel price. State regulators have an incentive to approve or encourage this type of cost allocation because it prevents incremental gas costs from being allocated to any party in its jurisdiction. Residential customers pay less of the fixed cost of distribution while industrials pay no more than they must under incremental pricing.

Implications

The downstream regulatory regime is itself likely to be subjected to new stresses by field price deregulation. The carrier status of pipelines may need to change, as uncertainties introduced by deregulation appear to justify a shift toward contract carriage and away from private carriage. To assess the relative merits of different types of carrier status, it is necessary to understand better the economic functions served by pipelines as private carriers, and how those functions would be carried out under alternative modes of operation. The regulatory status of gas pipelines may also change, either in the direction of more stringent regulation under common carrier rules or toward deregulation. In addition to the issues related to choice of carrier status, three additional questions arise in regard to changes in regulatory status:

How common carrier rules would be set and enforced in practice

How the results of common carrier status would differ from the results of contract carriage under the current FERC practice of rate-of-return regulation

Whether competition among pipelines in downstream markets is sufficient to protect consumers in a deregulated environment

Changes within the current regulatory regime may also be seen. Neither the formulas used for allocating fixed costs between demand and commodity charges nor incremental pricing rules allow efficient allocation of gas among customer classes and times of use. If deregulation of field prices is to provide maximum benefits to consumers, reconsideration of these rules is required. Such reconsideration requires a synthesis of optimal pricing rules for regulated utilities.

7

A Research Agenda

Introduction

The Natural Gas Policy Act (NGPA) of 1978 set the nation on a course toward partial deregulation of natural gas prices by 1985, but that deregulation applies only to the first sale of gas by producers to pipelines. Interstate pipelines remain under the jurisdiction of the Federal Energy Regulatory Commission (FERC), and the distribution companies to which pipelines sell gas remain regulated by state and local governments. The entire system—producers, pipelines, and distributors—remains knitted together by long-term contracts, the provisions of which are only slowly changing in response to the new environment created by wellhead price deregulation. Moreover, the high fixed cost of pipeline construction and economies of scale in distribution create conditions favorable to monopolistic behavior in various parts of the industry. As a result, field price deregulation, whether under the NGPA schedule or in a more complete and rapid transition, will not be the end of the story of problems in natural gas markets. The combination of regulation, rigidities introduced by long-term contracts, and market power means that price signals will not be transmitted without distortion between producers and consumers. Demand and supply may still fail to achieve equilibrium. Increases in gas costs arising from deregulation may be allocated inefficiently or inequitably across consumers, and the balance between risks and rewards will change in various segments of the industry.

Toward an Integrated Analysis of Gas Markets

With effective, if not complete, deregulation of natural gas prices at the field level, attention must turn to the remaining downstream portion of the natural gas industry. The three major institutional features of downstream markets interact to determine their performance and problems:

The tradition of long-term contracting for gas supplies, which tends to create price inflexibility and rigidity in response to new market conditions

Market power, at the distribution company

level and pipeline level, which may create inefficient or inequitable pricing and distribution patterns

Regulation, by state commission and the FERC, which affects price structures, rates of return, and investment decisions of pipelines and distributors

The maxim that in economics everything depends on everything else, in at least two ways, perhaps understates the complexity of gas market institutions and their consequences. By centering attention on pricing and allocation of gas at the burner tip and moving the focus of analysis farther and farther upstream toward the wellhead, it is possible at least to keep the major features in sight and to organize investigation of their relationships.

An analysis of monopolistic price discrimination under realistic regulatory constraint will yield understanding of burner tip pricing. To be useful, such analysis should include a representation of the process of state commission regulation—how decisions are made, what influences them, and how regulated firms are controlled—along the lines pioneered by Joskow. This process can be expected to change under the influence of new market conditions, so that alternative assumptions should be made in formal modelling of the effects of regulation on downstream behavior. Economic factors affecting distribution utility behavior, such as demand elasticities, ratios of peak to off-peak demand, and market structure—including interfuel and intercompany competition, access to storage, and presence of rival pipelines—will likewise vary across jurisdictions and must be investigated empirically.

The theoretical model of a distributor should encompass standard possibilities of price discrimination (time of use, inverse elasticity) and assignment of service reliability to various classes of customers, all under realistic regulatory constraint. Predictions of rate structures, capacity choices, and storage levels for different distribution companies could be derived from such a model and tested against observations. If there is sufficient variability among regional submarkets, the effects of differing regulatory regimes and demand conditions could be estimated statistically. An evaluative standard, based on usual concepts of Pareto optimality, must also be developed.

Links to the upstream gas market come through the distributors' derived demand for gas, which will vary over time and reflect the uncertainty of demand at the burner tip. The nature of transactions between distributors and pipelines must be studied, particularly in regard to the effects of regulation on the derived demand for gas, long-term contracting, and distinction between paying for reliability and paying for delivery. The terms on which gas is sold by pipelines to distribution companies will be determined simultaneously by the behavior of pipelines and distributors. The distributors' demand elasticity, and the variation of demand seasonally and stochastically, will be a product of choices the distributor makes, conditioned on characteristics of its customers' demand, storage possibilities, and its own regulatory regime. The degree of competition in the particular market area will also matter. This must be matched to the pipelines' pricing and supply decisions.

A model of pipeline behavior must encompass the regulatory, contracting, and market power issues mentioned. The process of FERC regulation should be examined and the current process and alternatives modeled analogously to state regulation. Through application of the methods of principal-agent theory, modelling of contractual relationships with producers and distributors will be possible. Combined with investigations of competitive structure, these developments will make possible quantitative analysis of pipeline gas purchasing behavior, investment, pricing of jurisdictional gas sales and sales to direct users, and decisions about reliability of supply. Because of the variability of economic and cost factors, a continued case study moving back from selected distributors to pipelines serving them may be useful.

To close the model, producer supply behavior must be brought in, with attention to the exhaustible nature of the resource and the nature of fixed investments and risks involved. In this analytical framework, it would be possible to trace out the effects of events and policies from field to burner tip, accounting for market power, contractual constraints, and regulatory effects, and to identify those obstacles to efficient resource allocation that warrant policy intervention.

Given the regulatory constraints, the nature of transactional arrangements, and the degree of market power that may exist in natural gas markets, there is little reason to suppose that natural gas policy can be put to rest with field price deregulation. But there has been so little analysis of downstream market issues that guidance on further policy intervention is almost completely lacking. It is not clear where intervention is needed or how it should be designed.

Four areas of research are critical for building an adequate understanding of natural gas markets. They are:

How continued downstream regulation will affect the behavior of pipelines and distribution companies in the face of freer upstream markets, and how that regulatory regime may change in response to field price deregulation

What economic functions long-term gas contracts perform, and how the contractual relationship may adapt to the greater uncertainty introduced by field price deregulation

Why particular transactional arrangements—integration, contracting, spot purchases—do or do not appear in natural gas markets, and how the mix is likely to change

What determines the degree of competition among gas pipelines, and whether workable competition exists

Research on each question will not only fill a gap in current understanding of how downstream markets work, but will help resolve current or foreseeable policy issues. Taking each question in turn, the following four sections lay out a research design and indicate the policy implications of results that might be obtained.

Research Designs

Pipeline and Distribution Company Behavior Under Continued Regulatory Constraint

CHARACTERIZING DOWNSTREAM BEHAVIOR. No satisfactory theory or data base now exists on which to formulate predictions of how pipelines and distribution companies will behave under the influence of downstream regulation. It is also important to recognize that the regulatory process itself can change. Alternative ways of regulating should be explored, and their predicted consequences compared with the current regime. A representation of the incentives and constraints created by commission regulation can be developed along the lines of Joskow's research on electricity regulation. Such a theory should include assumptions about how commission decisions are made, how those decisions are affected by actions of the regulated firm, and what the constraints are that commissions impose on regulated firms. Predictions of the regulated firm's behavior should consider how its ability to affect commission actions alters the behavior of the firm.

On the basis of these assumptions, a formal theory of the behavior of a regulated firm can be built. For example, if patterns seen in electricity hold true for natural gas, the commission could be represented as determining a schedule of prices that must be charged by the regulated firm, possibly with provisions for passing through some cost increases. Price schedules would be adjusted at periodic rate reviews, which occur when profits (or other performance indicators) reach either some upper or lower bound. New price schedules would be based on cost and performance data which can be altered by actions of the regulated firm. The firm can be represented as maximizing profits subject to these constraints. This approach to modelling regulatory constraints and incentives can be applied to both state commissions (relevant to distributors and intrastate pipelines) and to the FERC (relevant to interstate pipelines).

Also entering into the regulated firm's calculation will be parameters describing the economic environment in which it buys and sells gas. For distributors, these include demand elasticities of various customer classes and seasonal and stochastic influences on gas use—including those which determine the nature and mix of peak versus off-peak demand. Because these factors are likely to vary among regional markets, case studies are likely to be the most efficient way to proceed. Pipeline behavior will depend on elasticities of demand in regulated and unregulated markets, on the degree of rivalry with other pipelines, and on the mix of peak and off-peak demand. Opportunities to store gas may be exploited by either pipelines or distributors, and the different availability of storage across submarkets may produce systematic differences in pricing of peak and off-peak service.

The transaction between pipelines and distributors would be particularly important in this theory. The transaction is interesting because each party is separately subject to regulation, and each may possess some market power. The distributor's derived demand for gas, given the effects of state regulation and any market power the distributor can exercise, is unlikely to mirror faithfully its customers' willingness to pay. The pipeline's demand for gas is derived from the distributors, and subject to similar distortions. The result is likely to be that effective demand in field markets bears little relation to the preferences of ultimate consumers. Only careful study of actual regulatory processes can identify opportunities for real improvement.

Transactional arrangements between distributors and pipelines are more complex than the simple spot transactions typically considered in analyses of regulation. Below we lay out a general research design on transactional arrangements, but it is worth noting here that the nature of long-term contracting and the possibility of vertical integration by distributors should be analyzed in light of the incentives and constraints created by regulation. The converse is also true, that analysis of how behavior is conditioned by regulation must be done in conjunction with analysis of the determinants of transactional arrangements.

PRICING GAS TO END USERS. The combination of formal modelling of the regulatory process with empirical research on the economic environment in which particular pipelines and distributors operate will provide new tools for evaluating potential changes in regulatory practice.

One of the more important potential changes has to do with how cost increases will be allocated—between peak and off-peak sales, among retail sales and to various classes of customers, and between regulated sales at retail and unregulated direct sales to large industrial customers. These questions should have priority because of the likelihood of legislative and regulatory action to deal with perceived problems of how gas price increases under NGPA are allocated among final users. Changes in the incremental pricing formula of NGPA and in the formulas used to allocate fixed costs among demand and commodity charges are concrete examples of such action but by no means exhaust the possibilities.

Research should be directed toward a synthesis of the institutional and economics literature on cost allocation and two-part tariffs. A fruitful path for inquiry may be the development of a model of optimal gas utility pricing which harmonizes the results of the peak-load pricing literature, the "value of service" (inverse elasticity) pricing literature, and cost allocation literature.

For evaluating the effects of changes in the regulatory process and market outcomes, it is necessary to develop a normative standard. Conditions that would support an optimal allocation of gas among customers and times of use and an efficient combination of storage and transmission capacity must be identified and compared with the conditions created by current regulatory practice to determine if proposed regulatory reforms are worthwhile.

Efficient prices are based on the marginal cost of providing service (including capacity to meet peaks), but prices may instead be based on regulatory status of the sale, demand elasticities of various classes of customers, and rate base effects of peak period sales. Regulated monopolies that also do business in unregulated markets have an incentive to charge joint fixed costs to regulated sales, thus increasing their rate base and allowed profits, and to sell at less than long-run marginal cost in unregulated markets to undercut competition. This could include pricing gas at competitive fuel prices in direct industrial sales, even if such prices do not cover marginal costs of gas and transmission capacity. Departures from marginal cost pricing based on elasticity of demand will be adopted by a monopolist to maximize profits, even under rate-of-return regulation, although such departures can also be justified in a decreasing cost industry to allow the utility to break even. Two-part tariffs can increase revenues to a break-even level more efficiently, but such tariffs can also be a means of extracting monopoly rents. Such tariffs may not be sustainable in the face of competition.

Theoretical and empirical research is necessary to identify the welfare implications of different pricing methods. Case studies are required to examine the particular patterns of demand elasticities, peak/off-peak differentials, and storage opportunities available to different systems. Different practices among state regulatory authorities will also create different incentives for retail utilities and different pricing results. Behavior of regulatory agencies can be modelled formally to provide realistic representation of regulatory constraint. Predictions of pricing patterns under alternative economic and regulatory environments can be tested against actual outcomes. Theoretical work is also needed to provide an integrated characterization of optimal pricing patterns that take into account period of use, break-even constraints, and marginal cost of different types of service.

PIPELINE GAS PURCHASES. Changes in downstream regulation to deal with ways in which pipelines purchase gas may also be contemplated. Early analyses of the consequences of NGPA for pipeline behavior concentrated on the incentives to "bid-up" the price of some categories of gas above levels that would be established in a completely deregulated field market. In 1982, attention turned to questions of the incentives that pipelines have to choose the least expensive sources of gas, and proposals were made to strengthen those incentives by restricting the ability of pipelines to pass costs on to distributors. Some alternatives considered are restricting automatic operation of purchased gas adjustments, making a pipeline's return on assets conditional on performance in minimizing costs or maximizing sales, and abrogating minimum bill provisions of contracts. Evaluating the incentives of pipelines when they purchase gas requires a formal model of the regulatory process, under current and alternative procedures, such as that described. How pipelines will behave depends on how regulation works and, in particular, on the length of regulatory lag and on how much discretion pipelines actually have in constructing resale price schedules.

Changes in regulation that alter the lag between cost increases and rate adjustments must also be evaluated in light of their effects on the financial condition of gas utilities. As has been seen in the experience of electric utilities in the 1970s, regulatory lag in a period of rising input costs can make revenues lag permanently behind costs. Continued use of long-term take or pay contracts may also continually ratchet expenditures upward, as pipelines sign long-term contracts for reserves to meet short-run increases in demand, and then find that the higher costs drive future demand down below contracted levels of supply.

CARRIER STATUS. A more drastic change in regulation would be one that altered the basic carrier status of pipelines. Proposals to impose common carrier status on pipelines have come from some small producers who are expressing fears of pipeline monopoly of particular routes. Some pipeline customers have also expressed dissatisfaction with pipeline purchasing practices. Pipelines themselves have become increasingly uncomfortable with buying gas on long-term contract for resale in a much more volatile market. Private, contract, and common carrier status each assign to pipelines a different mix of economic functions, and thus provide producers, pipelines, and customers with a different mixture of risks and returns. These functions include transportation, a merchant function of buying and selling gas, brokerage, providing storage, and allocating pipeline capacity among peak users.

After the distribution and performance of these functions in different types of carrier status are analyzed, the relative merits of alternative carrier status for pipelines and of likely directions of change can be assessed. Such an assessment can provide a basis for deciding whether the current problems are being addressed by legislative or regulatory action or by allowing carrier status to evolve on its own. One distinguishing feature of the various carrier statuses is the regulatory process by which rates are set and access provided. The form of such processes and the outcomes they might produce could also be explored in light of the experience of other transport industries subject to such regulation.

PRICE AND SECURITY OF INTERRUPTIBLE SUPPLY. Finally, research on regulatory and contractual methods to match supply and demand for reliability would be important. Innovative methods of pricing interruptible service and of compensating producers for holding gas in the ground may be required if deregulated supply and uncertain demand are to be matched efficiently. Changes in downstream regulation would be required to meet this goal. Prices paid by ultimate consumers and producers can be structured to provide an incentive-compatible revelation of consumer preferences and payment for reserves and delivery. In particular, two-part pricing to producers and to interruptible customers may be necessary to reveal the importance attached to reliable service. In this way, reliability of service could be contracted for separately from payments for gas delivered.

Current provisions of long-term gas supply contracts and downstream regulatory practice are not likely to produce efficient levels of reserve additions, deliverability, or line-pipe and storage capacity after field price deregulation. The degree of interruptibility supplied may differ unnecessarily from that desired by consumers, and pipelines may purchase more or less reserves than would most efficiently meet demand. The nature of downstream regulation distorts a pipeline's incentives to provide peak and off-peak service. Conflicting incentives exist, some in the direction of excessive reliability and some in the direction of inadequate reliability. Regulation may also distort the choices pipelines and distribution companies make between storage and transmission line capacity.

Contracts between pipelines and producers fail to incorporate mechanisms to compensate producers for holding reserves and excess production capacity behind the pipe, with the result that production is not adjusted efficiently to meet variable demand.

Fundamental facts about the responsiveness of gas demand to reliability need to be established. Econometric studies can relate interruptible customers' demand to expected reliability. System optimization studies should find jointly optimal combinations of reserves, production, transmission and storage capacity, and match the minimum cost of providing given reliability levels with willingness to pay. To assess current incentives that affect reliability, representative contracts between producers and pipelines and between pipelines and interruptible customers should be analyzed intensively and compared with ideal pricing schemes. Research should also be directed at dynamic pricing models that can describe price paths that provide an adequate return to producers and a means of clearing markets during transitory demand shifts. The role of shorter term contracts, spot purchases, and off-system sales in matching supply and demand, and the contribution of PGAs should be examined.

The Economic Functions of Long-Term Contracts

Long-term contracts are the main vehicle through which transactions between producers and pipelines and between pipelines and distributors are carried out in the gas industry. Because of their continuing importance, research should be directed toward understanding what kinds of contractual arrangements provide the greatest benefits, first to the parties involved and second, to the economy as a whole.

The economic functions of specific contract structures can be elucidated on the basis of insights developed in the economic theory of agency and in studies of optimal contracts. Basic theoretical research is required to extend the principal-agent approach to incorporate uncertainty in the needs of the principal as well as in the effects of the agent's actions and to allow for the possibility that the principal as well as the agent can take actions that affect the utility of both. The empirical basis for applying such a theory appears to have been established through surveys of producer-pipeline contracts (EIA, 1982a).

The first step in applying the theory is to characterize the risks and incentives created by specific contract terms and to predict behavior of each party, taking contract terms as given. The second step is to represent contract terms as a matter of negotiation between the parties, and the negotiation process as one which leads to selection of Pareto-optimal contract terms, under realistic constraints on the amount of information available to each party. This representation can be used to predict under what conditions certain provisions will be adopted by mutual consent of the parties.

Finally, a "first-best optimum" that could be achieved if the actions of both parties were centrally directed can be defined, and the predicted outcome of the negotiation process compared with this evaluative standard.

The following steps would be required to make a scientifically valid prediction and assessment of contract terms. An explicit, formal model of natural gas contracting is required as a starting point.

1. Assign values to variables describing the interstate market before NGPA.
2. Predict pre-NGPA interstate contract terms.
3. Test predictions against representative contracts.
4. Repeat 1–3 for pre-NGPA intrastate contacts.
5. Accept or reject model.
6. Assign value to variables in accepted model to describe deregulated market.
7. Predict contract terms under deregulation.
8. Compute "optimal" outcomes and compare with equilibrium.

The greatest difficulties are likely to be in representing the space of possible contract terms in a realistic and mathematically tractable way, and in obtaining meaningful measures of the costs of more complex contracts and greater monitoring.

Contractual relations between regulated pipelines and regulated distributors also shift risks and create incentives, and can be studied by applying the principal-agent approach. It is impossible to understand the effects of regulation on pipeline and distribution behavior, and on the efficiency with which final consumers are served, without analyzing the contracting process. Further, changes in regulation also affect risks and incentives and thus may alter the form of contractual arrangements. Similar considerations apply in the upstream transaction with producers, although commission regulation affects only one side of that transaction directly. However, downstream regulatory changes will still influence the contractual process.

Certain provisions typically found in contracts between producers and pipelines have been blamed for causing perverse developments in natural gas markets. These include take-or-pay provisions that have been alleged to force pipelines to take high-cost gas and cut back purchases of low-cost gas, and indefinite price escalators that may drive prices well above a level that would clear the market. Minimum bill requirements affecting distributors have also become a concern.

Several legislative proposals to abrogate or limit the operation of such provisions have been made. The results of the research design sketched out can be used to evaluate the consequences of such actions, since an understanding is required of what economic function those provisions serve, of how they relate to other provisions that are or might be included in contracts, and of how changing provisions would affect the incentives and risks that face parties to the contract. Without this understanding, intervention in the contracting process may be ineffectual or may trigger changes worse than those the intervention is designed to correct. The analysis can also identify provisions that are dysfunctional in the current market, and suggest alternatives that could be beneficial to all parties.

The Choice of Transactional Arrangements

As the previous section notes, the new sources of price uncertainty introduced by field price deregulation will likely make long-term contracting a more costly and less efficient means of arranging transactions. A Pareto-optimal contract must involve some sharing of risk between the parties, assuming both are averse to risk. As price uncertainty increases, sharing of risks will require either more complex contracts, with associated costs of negotiation or marketing, or less effective incentives for effort in production and monitoring. As a result, greater inefficiencies in resource allocation under long-term contracts can be expected to appear. Not only will there likely be alterations in contract terms, but parties will have reason to search for alternative transaction modes. There could be a shift away from long-term contracts toward more integration, toward more "spot" sales, or toward both.

A fruitful way of organizing research on the choice of transactional arrangements that will emerge under upstream deregulation is through a comparative analysis of the transaction costs of contractual arrangements, vertical integration, and market-based institutions. There are two broad questions that such a research design ought to address: (1) How will upstream deregulation alter transactional arrangements among the industry's three vertical segments? and (2) Will the outcomes be socially optimal, that is, what role is there for public policy? Answering the first question requires a research design that is predictive; the second requires an evaluative analytical framework.

It is possible that for reasons of economic efficiency or because of institutional/regulatory constraints, nonmarket long-term arrangements will dominate the structure of transactions in gas. Moreover, there will continue to be a *mix* of contractual and integrated arrangements. But how will NGPA alter the mix of transactional arrangements?

The literature on the determinants of transactional arrangements is primarily cross-sectional: It provides a vehicle for understanding why transactions are structured differently across industries. Like most industries throughout the economy, firms within the natural gas industry—more precisely firms within each vertical segment of the industry—are not homogeneous. Pipelines differ from one another in important dimensions such as site, regional market/customer diversification, market share, and so on. These dimensions are likely to account for differences in their exposure to and ability to overcome transaction costs, as well as shape their capacity to respond to changes in risks and uncertainty. Differences also exist among producers and distributors. Under NGPA this heterogeneity will increase. Thus the conceptual apparatus offered in the literature can be used to explain why, at a given time, we might expect certain groups of firms within a particular vertical segment of the natural gas industry to structure a larger (or smaller) proportion of their post-NGPA transactions through contracts.

However, the tools from the literature are not directly applicable to analyze *dynamic* changes in transactional arrangements. It is in this area that basic research is needed to predict accurately how NGPA will change these arrangements. Most important, an analytical framework is needed that can be applied to nonrenewable resources. The simple models in the literature cannot adequately capture how deregulation of an exhaustible resource will affect the relative cost of various transactional arrangements over time.

To complete the answer to the first question, analysis of the prospects for the emergence of new market-based transactional arrangements is required. Natural gas will continue to be produced inefficiently even with field deregulation because of contractual arrangements that insulate suppliers from changes in gas demand. In general, contract terms provide incentives for effort and to divide risks among the parties. Immediate netback of retail prices to the producer creates incentives for efficient gas production but places all risks of price change on the producer. On average, higher prices will be required to compensate producers for bearing such risks than would be required with fixed escalators that place price risks with the pipeline. If alternative sources of insurance or hedging, such as futures markets, could be developed, efficiency in production and in allocation of risk would be reconciled. However, futures markets cannot exist without spot markets, and the transportation costs and specialized capital characteristic of the gas market may make a pervasive spot market impossible.

Research should begin by examining capital equipment involved in the production and gathering of natural gas, and the nature of competition among producers to assess the importance of economic barriers to spot transactions. The chances that spot prices will be sufficiently uniform, geographically and across transactions, to support futures markets should be examined. Whether the proportion of futures to spot transactions will be enough to provide adequate hedging should be investigated using agricultural and other commodity markets as a yardstick. Actual contract terms should be evaluated according to their ability to allocate gas efficiently and their risk-sharing properties. Rough numerical estimates of social surplus losses from rigid contract terms can be made.

The second question to be answered through future research concerns the impact on economic welfare of whatever transactional arrangements appear. To be sure, the private incentives created by NGPA in this area may bring about socially undesirable outcomes. Downstream firms will remain under utility-type regulation, and it is by now well known that firm behavior under this kind of regulatory constraint may diverge from the social optimum. Moreover, the partial deregulation embodied in NGPA, unless it is scrapped in favor of complete field price decontrol, itself introduces distortionary effects on private sector behavior away from the social optimum. In short, even at the outset of this era of natural gas deregulation, we are far away from a first-best world. This is important because signals that might otherwise suggest appropriate public sector intervention may actually conceal a situation that requires a different, and possibly opposite, role for public policy.

With these caveats, the literature does provide reasonably strong tools to judge the welfare implications of the mix of transactional arrangements that will emerge under NGPA. For example, if firms operating in relatively concentrated markets—that is, markets in which the degree of horizontal dominance is substantial—are found to integrate vertically, the net impact on economic welfare could be negative even if transaction costs are reduced in the process. In this instance, public policy can play an appropriate role.

Interpipeline Competition

The state of competition among pipelines is a critical piece of information to be considered in reforming pipeline regulation. In addition, competitive issues arise in analyzing the consequences of field price deregulation. How pipelines behave under regulatory constraint depends on their market power, as does the evaluation of alternatives involving greater regulatory stringency (such as common carrier regulation) or deregulation of pipelines sales. NGPA may directly affect market shares as a result of differentials in cheap gas cushions among pipelines, and these changes in market shares may increase concentration further in some pipeline markets that are already relatively concentrated.

Assessing the state of competition requires an approach that integrates appropriate structural measures of concentration with analysis of other elements that determine competitive conditions in pipeline markets and that may make exercise of market power more or less difficult. These other elements include:

How might an increase in vertical integration affect interpipeline competition?

How do transportation distances, capital barriers to entry, and the availability of substitute fuels combine to determine the degree of competition in local receiving and delivery markets?

Does regulation of distribution make easier or harder the exercise of any structural market power that pipelines might possess?

Are strategic options such as entry-deterring investment and limit pricing open to pipelines that can be used to maintain market power?

How might the structural conditions of pipeline markets affect the competitive position of natural gas at the burner tip?

How do contractual arrangements with distributors affect the scope of competition?

The theoretical literature in industrial organization provides the analytical tools with which judgments can be made about the structural condition of pipeline markets. However, there is relatively little recent empirical research that applies this analytical framework to the natural gas industry.

The first priority is to ascertain the criteria for geographic and product boundaries of pipeline markets and to develop theoretically satisfactory measures of market structure. Research should be directed toward a conceptual definition of the geographic boundaries for pipeline markets so horizontal structure can be measured.

In addition, research is needed to identify changes that NGPA per se might bring about in influencing the state of competition among pipelines. Partial deregulation of field prices under NGPA will cause substantial cost, and hence performance, differences among pipelines as a result of differences in cushions of cheap gas. A restructuring of the pipeline industry will likely take place with the prospect that some pipelines may achieve noncompetitive horizontal dominance in market shares across regions and across customer classes. To this end, a systematic data base on pipeline cushion differentials and other variables, for example, pipeline financial characteristics, market shares, etc., could be assembled and a statistical analysis could be performed to determine the extent to which cushion differentials contribute to increased concentration in pipeline markets.

Finally, this traditional approach to market power should be then combined with an assessment of the other elements that may affect interpipeline competition that were discussed above.

Conclusion

The influences on the gas market of downstream regulation, contracting practices, changes in transactional arrangements, and interpipeline competition are not adequately understood. Regulation and long-term contracts have particularly clear effects on the performance of the gas market, and analysis of these subjects has been largely neglected. Both theoretical and empirical research is required.

Perhaps even more important, however, is an understanding of how the various features of the market fit together. None works in isolation from the others, and neither the state of the gas market nor potential directions of change can be truly understood unless all the features are analyzed simultaneously. An analysis that works backward from demand conditions at the burner tip through the transactions at each vertical stage of the industry, and considers regulatory influence, transactional arrangements, and structural conditions at each step promises better understanding of those interactions.

Bibliography

American Gas Association. 1982. *Gas Facts, 1981 Data* (Washington, American Gas Association).

Areeda, Philip, and Donald F. Turner. 1975. "Predatory Pricing and Related Practices Under Section 2 of the Sherman Act," *Harvard Law Review* vol. 88 (February) pp. 697–733.

Arrow, Kenneth J. 1975. "Vertical Integration and Communication," *Bell Journal* vol. 6 (Spring) pp. 173–183.

Averch, Harvey, and Leland Johnson. 1962. "Behavior of the Firm under Regulatory Constraint," *American Economic Review* vol. 52 (December) pp. 1052–1069.

Bailey, Elizabeth E. 1972. "Peak-Load Pricing Under Regulatory Constraint," *Journal of Political Economy* vol. 80 (July) pp. 662–679.

———and Roger Coleman. 1971. "The Effect of Lagged Regulation in an Averch-Johnson Model," *Bell Journal* vol. 2 (Spring) pp. 278–292.

———and Eric B. Lindenberg. 1976. "Peak Load Pricing Principles: Past and Present" in Harry M. Trebing, ed. *New Dimensions in Public Utility Pricing* (East Lansing, Mich., Michigan State University) pp. 9–31.

———and John C. Malone. 1970. "Resource Allocation and the Regulated Firm," *Bell Journal* vol. 1 (Spring) pp. 129–142.

———and Lawrence White. 1974. "Reversals in Peak and Off-Peak Prices," *Bell Journal of Economics* vol. 5 (Spring) pp. 75–92.

Bain, Jose S. 1956. *Barriers to New Competition: Their Character and Consequences in Manufacturing Industries* (Cambridge, Mass., Harvard University Press).

Baumol, William, and David Bradford. 1970. "Optimal Departures from Marginal Cost Pricing," *American Economic Review* vol. 60 (June) pp. 265–283.

———and Alvin K. Klevorick. 1970. "Input Choices and Rate-of-Return Reg: An Overview of the Discussion," *Bell Journal* vol. 1 (Autumn) pp. 162–190.

———and C. Walton. 1973. "Full Costing, Competition and Regulatory Practice," *Yale Law Journal*, vol. 82, p. 639.

Blair, Roger, and David Kaserman. 1978. "Uncertainty and the Incentive for Vertical Integration," *Southern Economics Journal* vol. 46 (July) pp. 266–272.

Blaydon, Colin, Wesley Magat, and Celia Thomas. 1979. "Marginal Cost and Rate Structure Design for Retail Sales of Natural Gas." Michael Crew, ed. *Problems in Public Utility Economics and Regulation* (Lexington, Mass.: Lexington Books).

Borch, Karl. 1962. "Equilibrium in a Reinsurance Market," *Econometrica* vol. 30 (July) pp. 424–444.

Bork, Robert H. 1954. "Vertical Integration and the Sherman Act: The Legal History of an Economic

Misconception," *University of Chicago Law Review* vol. 22 (Autumn) pp. 157–201.

Braeutigam, Ronald R. 1979. "Optimal Pricing with Intermodal Competition," *American Economic Review* vol. 69 (March) pp. 38–49.

______. 1980. "An Analysis of Fully Distributed Cost Pricing in Regulated Industries," *Bell Journal* vol. 11 (Spring) pp. 182–196.

______. 1981. "The Deregulation of Natural Gas," Leonard W. Weiss and Michael W. Klass, eds. *Case Studies in Regulation: Revolution and Reform* (Boston, Mass., Little, Brown), pp. 142–186.

Breyer, Stephen. 1982. *Regulation and Its Reform* (Cambridge, Mass., Harvard University Press).

______ and Paul MacAvoy. 1974. *Energy Regulation by the FPC* (Washington, D.C., Brookings Institution).

Broadman, Harry G. 1981. "Intraindustry Structure, Integration Strategies and Petroleum Firm Performance" (Ph.D. dissertation, University of Michigan, Department of Economics).

______, W. David Montgomery, and Milton Russell, 1982. "Field Price Deregulation and Changes in the Regulatory/Carrier Status of Natural Gas Pipelines." Paper presented at the International Association of Energy Economists, Fourth Annual North American Meeting, Denver, Colorado (Washington, D.C., Resources for the Future).

Brown, Gardner, and M. Bruce Johnson. 1969. "Public Utility Pricing and Output Under Risk," *American Economic Review* vol. 59 (March) pp. 119–128.

Brown, Keith. 1970. *Regulation of the Natural Gas Producing Industry* (Baltimore, Md., Johns Hopkins University Press for Resources for the Future).

Burness, H. Stuart, W. David Montgomery, and James P. Quirk. 1980. "Capital Contracting and the Regulated Firm," *American Economic Review* vol. 70 (June) pp. 342–354.

Burstein, Meyer L. 1960. "A Theory of Full-Line Forcing," *Northwestern University Law Review* vol. 55, pp. 157–201.

Callen, Jeffrey L. 1978. "Production, Efficiency, and Welfare in the Natural Gas Transmission Industry," *American Economic Review* vol. 68 (June) pp. 311–323.

Camm, Frank A. 1978. *Average Cost Pricing of Natural Gas: A Problem and Three Policy Options* (Santa Monica, Rand Corporation).

Carlton, Dennis W. 1979. "Vertical Integration in Competitive Markets under Uncertainty," *Journal of Industrial Economics* vol. 28 (March) pp. 189–209.

Cheung, Steven. 1969. "Transaction Costs, Risk Aversion and the Choice of Contractual Arrangements," *Journal of Law and Economics* vol. 12 (April) pp. 23–42.

Cicchetti, Charles J., Philip J. Mause, and Rod Shaughnessy. 1979. "Role of the States in Implementing Pricing and Other Provisions of the National Energy Act" (Wisconsin Public Service Commission).

Coase, R. H. 1952. "The Nature of the Firm," *Econometrica,* vol. 4 (1937), reprinted in American Economic Association, *Readings in Price Theory* (Chicago, Irwin) pp. 331–351.

______. 1964. "The Regulated Industries—Discussion," *American Economic Review* vol. 54 (May) pp. 194–197.

Congressional Research Service. 1982. *Natural Gas Regulation Study* (Washington, Government Printing Office, July).

Corey, Gordon R. 1971. "The Averch and Johnson Proposition: A Critical Analysis," *Bell Journal* vol. 2 (Spring) pp. 358–373.

Cowan, Don, and Rich Hagar. 1982. "U.S. Gas Pipeline Firms Moving Deeper into Upstream Activity," *Oil & Gas Journal* (February) pp. 55–60.

Crawford, Vincent P. 1982. "Long-Term Relationships Governed by Short-Term Contracts," Discussion Paper no. 926 (Cambridge, Mass., Harvard University, Harvard Institute of Economic Research, November).

Crew, Michael A., and Paul R. Kleindorfer. 1976. "Peak-Load Pricing with a Diverse Technology," *Bell Journal* vol. 7 (Spring) pp. 207–231.

Cukierman, Alex, and Zalman F. Schiffer. 1976. "Contracting for Optimal Delivery Time," *Bell Journal of Economics* vol. 7 (Spring) pp. 132–149.

Cyert, R. M., and J. G. March. 1963. *A Behavioral Theory of the Firm* (Englewood Cliffs, N.J., Prentice-Hall).

Davidson, Ralph Kirby. 1955. *Price Discrimination in Selling Gas and Electricity* (Baltimore, Md., Johns Hopkins University Press).

Demsetz, Harold. 1973. "Joint Supply and Price Discrimination," *Journal of Law and Economics* vol. 16 (October) pp. 389–405.

______. 1968. "Why Regulate Utilities?" *Journal of Law and Economics* vol. 11, pp. 55–66.

Department of Energy, Office of Policy, Planning and Analysis (DOE/OPPA). 1981. "A Study of Alternatives to the Natural Gas Policy Act of 1978" (DOE-PE-0031) (Washington, D.C., Department of Energy, November).

DeVany, A. S., and T. R. Saving. 1977. "Product Quality, Uncertainty and Regulation: The Trucking Industry," *American Economic Review* vol. 67 (September) pp. 583–594.

Diamond, Peter A., and Eric Maskin. 1981. "An Equilibrium Analysis of Search and Breach of Contract: A Non-Steady State Example," *Journal of Economic Theory* vol. 25 (October) pp. 165–195.

———. 1979. "An Equilibrium Analysis of Search and Breach of Contract: Steady States," *Bell Journal* vol. 10 (Spring) pp. 282–316.

Ekelund, Robert B., Jr., and Robert F. Hebert. 1980., "Uncertainty, Contract Costs and Franchise Bidding," *Southern Economic Journal* vol. 47 (October) pp. 517–521.

——— and Richard S. Higgins. 1982. "Capital Fixity, Innovations and Long-Term Contracting: An Intertemporal Economic Theory," *American Economic Review* vol. 72 (March) pp. 32–46.

Energy Information Administration (EIA). 1981. "An Analysis of the Natural Gas Policy Act and Several Alternatives: Part I, The Current State of the Natural Gas Market" (Washington, D.C., Natural Gas Division, Office of Oil and Gas, Department of Energy, December).

———. 1982a. "Natural Gas Producer/Purchaser Contracts and Their Potential Impacts on the Natural Gas Market" (DOE-EIA-0330) (Washington, D.C., Department of Energy, June).

——— 1982b. "An Analysis of Post-NGPA Interstate Pipeline Wellhead Purchases" (DOE-EIA-0357) (Washington, Department of Energy, September)

——— 1982c. "Natural Gas Monthly" (DOE/EIA-0130[82/12]) (Washington, Department of Energy, December).

———. 1983. "Natural Gas Monthly" (DOE/EIA-0130[83/1]) (Washington, Department of Energy, January).

Erickson, Edward W., and Robert M. Spann. 1971. "Supply Response in a Regulated Industry: The Case of Natural Gas," *Bell Journal* vol. 2 (Spring) pp. 94–121.

Faulhaber, Gerald R. 1975. "Cross-Subsidization: Pricing in Public Enterprises," *American Economic Review* vol. 65 (December) pp. 966–977.

Federal Energy Regulatory Commission (FERC), Office of Pipeline and Producer Regulation. 1981. "Discussion Paper on Gas Pipeline Rate-Making Draft for Internal Distribution" (Washington, December 4).

———. Office of Regulatory Analysis. 1982. "Off-System Sales: A Preliminary Outline of the Policy Issues" (October 14).

Flaherty, M. Therese. 1981. "Prices versus Quantities and Vertical Financial Integration," *Bell Journal* vol. 12 (Autumn) pp. 507–525.

Friedlaender, Ann F. 1969. *The Dilemma of Freight Transportation Regulation* (Washington, D.C., Brookings Institution).

Garfield and Lovejoy. 1964. *Public Utility Economics* (Englewood Cliffs, N.J., Prentice-Hall).

Goldberg, Victor P. 1976a. "Toward an Expanded Economic Theory of Contract," *Journal of Economic Issues* vol. 10 (March) pp. 45–61.

———. 1976b. "Regulation and Administered Contracts," *Bell Journal* vol. 7 (Autumn) pp. 426–448.

Gravelle, H. S. E. 1976. "The Peak-Load Problem with Feasible Storage," *The Economic Journal* vol. 86 (June) pp. 256–277.

Green, Jerry R. 1974. "Vertical Integration and Assurance of Markets." Discussion Paper no. 383 (Cambridge, Mass., Harvard Institute of Economic Research).

———. 1982. "Statistical Division Theory Requiring Incentives for Information Transfer," in John J. McCall, ed. *The Economics of Information and Uncertainty* (Chicago, Ill., University of Chicago Press).

Greenhut, M. L., and H. Ohta. 1976. "Related Market Conditions and Interindustrial Mergers," *American Economic Review* vol. 66 (June) pp. 267–277.

Haring, John R., and David L. Kaserman. 1978. "Related Market Conditions and Interindustrial Mergers: Comment," *American Economic Review* vol. 68 (March) pp. 225–227.

Harris, Milton, and Artur Raviv. 1979. "Optimal Incentive Contracts with Imperfect Information," *Journal of Economic Theory* vol. 20 (April) pp. 231–259.

Harris, Richard. 1981. "Price and Entry Regulations with Large Fixed Costs," *Quarterly Journal of Economics* vol. 96 (November) pp. 643–655.

Hay, George A. 1973. "An Economic Analysis of Vertical Integration," *Industrial Organization Review* vol. 1, pp. 188–198.

Hayashi, Paul M., and John M. Trapani. 1976. "Rate of Return Regulation and the Regulated Firm's Choice of Capital-Labor Ratio: Further Empirical Evidence on the Averch-Johnson Model," *Southern Economic Journal* vol. 42 (January) pp. 384–397.

Hirshleifer, Jack. 1958. "Peak Loads and Efficient

Pricing: Comment," *Quarterly Journal of Economics* vol. 72 (August) pp. 451–462.

______ and John G. Riley. 1979. "The Analysis of Uncertainty and Information—An Expository Survey," *Journal of Economic Literature* vol. 17 (December) pp. 1375–1421.

Holstrom, Bengt. 1979. "Moral Hazard and Observability," *Bell Journal* vol. 10 (Spring) pp. 74–91.

Hotelling, Harold. 1931. "The Economics of Exhaustible Resources," *Journal of Political Economy* vol. 39 (April) pp. 137–175.

Houthakker, H. S. 1959. "Scope and Limits of Futures Trading," in Moses Abramovitz et al., *The Allocation of Economic Resources: Essays in Honor of Bernard Haley* (Stanford, Calif.).

Hudson, Edward A. 1982. "Energy, Productivity and Economic Effects of Energy Price Decontrol." MIT Studies in Energy and the American Economy, Discussion Paper No. 19 (Cambridge, Mass., Massachusetts Institute of Technology).

Ingram, William Coleman, 1976. "The Effect of State Regulation on Natural Gas Utilities" (Ph.D. dissertation, West Virginia University).

Isaac, R. Mark. 1982. "Fuel Cost Adjustment Mechanisms and the Regulated Utility Facing uncertain Fuel Prices," *Bell Journal* vol. 13 (Spring) pp. 158–169.

Joskow, Paul L. 1972. "The Determination of the Allowed Rate of Return in a Formal Regulatory Hearing," *Bell Journal* vol. 3 (Autumn) pp. 632–644.

______. 1973. "Pricing Decisions of Regulated Firms: A Behavioral Approach," *Bell Journal* vol. 4 (Spring) pp. 118–140.

______. 1974. "Inflation and Environmental Concern: Structural Change in the Process of Public Utility Price Regulation," *Journal of Law and Economics* vol. 17 (October) pp. 291–327.

______ and Roger Noll. 1973. "Relative Prices on Regulated Transactions of the Natural Gas Pipelines," *Bell Journal* vol. 4 (Spring) pp. 212–234.

______ and Roger Noll. 1980. "Theory and Practice in Public Regulation: A Current Overview," Conference Paper No. 64 (New York, National Bureau of Economic Research).

Kahn, Alfred E. 1961. "The Chemical Industry," in Walter Adams, ed. *The Structure of American Industry*, 3d ed. (New York, Macmillan).

______. 1970, 1971. *The Economics of Regulation*, 2 vols. (New York, Wiley).

Klein, Benjamin, Robert G. Crawford, and Armen A. Alchian. 1978. "Vertical Integration, Appropriable Rents, and the Competitive Contracting Process," *Journal of Law and Economics* vol. 21 (October) pp. 297–326.

______ and Keith B. Leffler. 1981. "The Role of Market Forces in Assuring Contractual Performance," *Journal of Political Economy* vol. 89 (August) pp. 615–641.

Klevorick, Alvin K. 1971. "The 'Optimal' Fair Rate of Return," *Bell Journal* vol. 2 (Spring) pp. 122–153.

Lax, David A., and James K. Sebenius. 1981. "Insecure Contracts and Resource Development," *Public Policy* vol. 29 (Fall) pp. 417–436.

Lee, Dwight R. 1978. "Price Controls, Binding Constraints and Intertemporal Economic Decision-Making," *Journal of Political Economy* vol. 86 (April) pp. 293–301.

Little, James D. 1971. "The Atlantic Seaboard Formula and Pipeline Rate Design," in Harry M. Trebing, ed. *Essays on Public Utility Pricing and Regulation* (East Lansing, Michigan State University) pp. 333–340.

Littlechild, Stephen C. 1970a. "A Game-Theoretic Approach to Public Utility Pricing," *Western Economic Journal* vol. 8 (June) pp. 162–166.

______. 1970b. "Marginal-Cost Pricing with Internal Costs," *The Economic Journal* vol. 80 (June) pp. 323–335.

______. 1975a. "Common Costs, Fixed Charges, Clubs and Games," *Review of Economic Studies* vol. 42 (January) pp. 117–124.

______. 1975b. "Two-Part Tariffs and Consumption Externalities," *Bell Journal* vol. 6 (Autumn) pp. 661–670.

MacAvoy, Paul. 1953. *Price Formation in Natural Gas Fields* (New Haven, Yale University Press).

______ and Roger Noll. 1973. "Relative Prices on Regulated Transactions of the Natural Gas Pipelines," *Bell Journal* vol. 4 (Spring) pp. 212–234.

______ and Robert Pindyck. 1973. "Alternative Regulatory Policies for Dealing with Natural Gas Storage," *Bell Journal* vol. 4 (Autumn) pp. 454–498.

McKie, James W. 1970. "Regulation and the Free Market: the Problem of Boundaries," *Bell Journal* vol. 1 (Spring) pp. 6–26.

McNicol, David L. 1973. "The Comparative Static Properties of the Theory of the Regulated Firm," *Bell Journal* vol. 4 (Autumn) pp. 428–453.

Machlup, Fritz, and Martha Tauber. 1960. "Bilateral Monopoly, Successive Monopoly, and Vertical Integration," *Economica* vol. 27 (May) pp. 101–119.

Marino, Anthony M. 1981. "Optimal Departures

from Marginal Cost Pricing: The Case of a Rate of Return Constraint," *Southern Economic Journal* vol. 48 (July) pp. 37–49.

Marshall, John. 1974. "Insurance Theory: Reserves as Mutuality," *Economic Inquiry* vol. 12 (December) pp. 476–492.

Marshall, John M. 1974. "Private Incentives and Public Information," *American Economic Review* vol. 64 (June) pp. 373–390.

———. 1976. "Moral Hazard," *American Economic Review* vol. 66 (December) pp. 880–890.

Mead, David E. 1981. "Concentration in the Natural Gas Pipeline Industry." Preliminary draft (Washington, D.C., Office of Regulatory Analysis, Federal Energy Regulatory Commission).

Means, Robert C. 1982. "Issues in the Debate Over Natural Gas Decontrol," *Public Utilities Fortnightly* (October 28) pp. 18–24.

Merrill, Peter R. "The Regulation and Deregulation of Natural Gas in the United States (1938–1985)" (Cambridge, Mass., Harvard University, Energy and Environmental Policy Center, Discussion Paper Series).

Meyer, Robert A. 1976. "Capital Structure and the Behavior of the Regulated Firm Under Uncertainty," *Southern Economic Journal* vol. 42 (April) pp. 600–609.

———. 1979. "Regulated Monopoly Under Uncertainty," *Southern Economic Journal* vol. 45 (April) pp. 1121–1129.

Mills, Gordon. 1976. "Public Utility Pricing for Joint Demand Involving a Durable Good," *Bell Journal* vol. 7 (Spring) pp. 299–307.

Mitchell, Edward J., ed. 1979. *Oil Pipelines and Public Policy: Analysis of Proposals for Industry Reform and Reorganization* (Washington, D.C., American Enterprise Institute).

Mulholland, Joseph P. 1979. *Economic Structure and Behavior in the Natural Gas Production Industry* (Washington, Federal Trade Commission).

Murrell, Peter. 1979. "The Performance of Multiperiod Managerial Incentive Schemes," *American Economic Review* vol. 69 (December) pp. 934–940.

Murry, Donald A. 1973. "Practical Economics of Public Utility Regulation: An Application to Pipelines" (with comment by Carl E. Horn), in Milton Russell, ed. *Perspectives in Public Regulation* (Carbondale, Southern Illinois University Press).

Needy, Charles W. 1976. "Social Cost of the A-J-W Output Distortion," *Southern Economic Journal* vol. 42 (January) pp. 486–489.

Nelson, J.R., ed. 1964. *Marginal Cost Pricing in Theory and Practice* (Englewood Cliffs, N.J., Prentice-Hall).

Neuner, E. J. 1960. *The Natural Gas Industry* (Norman, Okla., University of Oklahoma Press).

Newbery, David M. G., and Joseph E. Stiglitz. 1981. *The Theory of Commodity Price Stabilization* (Oxford, England, Clarendon Press).

Nguyen, D. T. 1976. "The Problems of Peak Loads and Inventories," *Bell Journal* vol. 7 (Spring) pp. 242–248.

Noll, Roger G., and Lewis A. Rivlin. 1973. "Regulating Prices in Competitive Markets," *Yale Law Journal*, vol. 82, pp. 1426–1434.

Oi, Walter Y. 1961. "The Desirability of Price Instability Under Perfect Competition," *Econometrica* vol. 29 (January) pp. 58–64.

Oil & Gas Journal. 1976. "U.S. Interstate Lines Pour Cash into Search for Gas" (December 27).

———. 1982. "U.S. Gas Pipeline Firms Moving Deeper into Upstream Activity" (February 8) pp. 55–59.

The Oil Daily. 1983 "Producers Fault AGA Study of Gas Purchase Costs" (March 15).

Okun, Arthur M. 1981. *Prices and Quantities: A Macroeconomic Analysis* (Washington, Brookings Institution).

Panzar, John C., and David S. Sibley. 1978. "Public Utility Pricing Under Risk: The Case of Self-Rationing," *American Economic Review* vol. 68 (December) pp. 888–895.

Peles, Yoram C., and Eyton Sheshinski. 1976. "Integration Effects of Firms Subject to Regulation," *Bell Journal of Economics* vol. 7 (Spring) pp. 308–313.

——— and Jerome L. Stein. 1976. "The Effect of Rate of Return Regulation Is Highly Sensitive to the Nature of Uncertainty," *American Economic Review* vol. 66 (June) pp. 278–289.

Perry, Martin K. 1978. "Related Market Conditions and Interindustrial Mergers: Comment," *American Economic Review* vol. 68 (March) pp. 221–224.

Pierce, Richard J., Jr. 1980. *Natural Gas Regulation Handbook* (New York, Executive Enterprises Publications).

Portman, Robert. 1982. "Competition Issues for Integrated Natural Gas Pipelines," Report to Division of Energy Deregulation (Washington, D.C., Department of Energy, October).

Posner, Richard A. 1971. "Taxation by Regulation," *Bell Journal* vol. 2 (Spring) pp. 22–50.

Pratt, Michael D. 1981. "Firm Behavior Under Regulatory Constraint: An Immanent Criticism," *Southern Economic Journal* vol. 48 (July) pp. 235–238.

Primeaux, Walter J., and Randy A. Nelson. 1980. "An Examination of Price Discrimination and Internal Subsidization by Electric Utilities," *Southern Economic Journal* vol. 47 (July) pp. 84–98.

Radner, Roy. 1970. "Problems in the Theory of Markets Under Uncertainty," *American Economic Review Papers and Proceedings* vol. 60 (May) pp. 454–460.

Roberts, R. Blaine. 1980. "Effects of Supply Contracts on the Output and Price of an Exhaustible Resource," *Quarterly Journal of Economics* vol. 95 (September) pp. 245–260.

Rosenberg, Laurence C. 1967. "Natural-Gas-Pipeline Rate Regulation: Marginal-Cost Pricing and the Zone Allocation Problem," *Journal of Political Economy* vol. 75 (April) pp. 159–168.

Ross, Stephen A. 1973. "The Economic Theory of Agency: The Principal's Problem," *American Economic Review Papers and Proceedings* vol. 63 (May) pp. 134–139.

______. 1979. "Equilibrium and Agency—Inadmissible Agents in the Public Agency Problem," *American Economic Review Papers and Proceedings* vol. 69 (May) pp. 308–312.

Rothschild, Michael, and Joseph E. Stiglitz. 1976. "Equilibrium in Competitive Insurance Markets: An Essay on the Economics of Imperfect Information," *Quarterly Journal of Economics* vol. 90 (November) pp. 630–649.

Russell, Milton. 1982. "Natural Gas Deregulation: Overview of Policy Issues." Discussion Paper D-92 (Washington, D.C., Resources for the Future, April).

______. 1983. "Gas Utilities and Market Clearing Prices: Adjusting to Fluctuations in Burner Tip Demand," speech before The Institute for Study of Regulation, Ninth Annual Rate Symposium (Washington, D.C., Resources for the Future, February 7).

Scheidell, John M. 1976. "Relevance of Demand Elasticity for Rate-of-Return Regulation," *Southern Economic Journal* vol. 43 (October) pp. 1088–1095.

Scherer, Frederick M. 1976. Comment on Areeda and Turner, *Harvard Law Review* vol. 89 (March) pp. 869–903.

______. 1980. *Industrial Market Structure and Economic Performance*, 2d ed. (Chicago, Rand McNally).

Schmalensee, Richard. 1973. "A Note on the Theory of Vertical Integration," *Journal of Political Economy* vol. 81 (March/April), pp. 442–449.

Schnabel, Morton. 1976. "Defining a Product," *Journal of Business* vol. 49 (October 1976) pp. 517–529.

Schwartz, David A., and John W. Wilson. 1974. Statement before the Senate Antitrust and Monopoly Subcommittee. *Hearings on the Natural Gas Industry*, June 27 (Washington, D.C., Government Printing Office).

Shavell, Steve. 1979. "Risk Sharing and Incentives in the Principal and Agent Relationship," *Bell Journal* vol. 10 (Spring) pp. 55–73.

Shepherd, William G. 1966. "Marginal-Cost Pricing in American Utilities," *Southern Economic Journal* vol. 33 (July) pp. 58–70.

______. 1972. "The Elements of Market Structure," *Review of Economics and Statistics* vol. 54 (February) pp. 25–38.

Sherman, Roger. 1981. "Pricing Inefficiencies under Profit Regulation," *Southern Economic Journal* vol. 48 (October) pp. 475–489.

______ and Michael Visscher. 1978. "Second Best Pricing with Stochastic Demand," *American Economic Review* vol. 68 (March) pp. 41–53.

Simon, H. 1959. "Theories of Decision-Making in Economics and Behavioral Science," *American Economic Review* vol. 49 (June) pp. 253–283.

Smiley, Robert. 1980. "A Marginalist Approach to Pricing U.S. Natural Gas," *Energy Economics* vol. 2 (July) pp. 172–178.

Spence, A. Michael. 1975a. "The Economics of Internal Organization: An Introduction," *Bell Journal* vol. 6, no. 1 (Spring).

______. 1975b. "Monopoly, Quality, and Regulation," *Bell Journal* vol. 6 (Autumn) pp. 417–429.

______. 1977. "Entry, Capacity, Investment and Oligopolistic Pricing," *Bell Journal* vol. 8 (Autumn) pp. 534–544.

______and Richard Zeckhauser. 1971. "Insurance, Information and Individual Action," *American Economic Review Papers and Proceedings* vol. 61 (May) pp. 380–387.

Steiner, Peter O. 1957. "Peak Loads and Efficient Pricing," *Quarterly Journal of Economics* vol. 71 (November) pp. 585–610.

Stigler, George J. 1971. "The Theory of Economic Regulation," *Bell Journal* vol. 2 (Spring) pp. 3–21.

Stollery, Kenneth R. 1981. "Price Controls on Non-Renewable Resources When Capacity Is Constrained," *Southern Economic Journal* vol. 48 (October) pp. 490–498.

Strand, Stephen H. 1980. "Regulatory Boundaries and Efficient Resource Allocations in the Production and Transmission of a Single Good," *Southern Economic Journal* vol. 46 (January) pp. 777–791.

Takayama, Akira. 1969. "Behavior of the Firm Under Regulatory Constraint," *American Economic Review* vol. 59 (June) pp. 255–260.

Townsend, Robert M. 1979. "Optimal Contracts and Competitive Markets with Costly State Verification," *Journal of Economic Theory* vol. 21 (October) pp. 265–293.

Tschirhart, John, and Frank Jen. 1979. "Behavior of a Monopoly Offering Interruptible Service," *Bell Journal* vol. 10 (Spring) pp. 244–258.

Tussing, Arlon R., and Connie C. Barlow. 1978. "An Introduction to the Gas Industry with Special Reference to the Proposed Alaska Highway Gas Pipeline (A Preliminary Report to the Alaska State Legislature)" (Anchorage, Alaska, University of Alaska, October 25).

———. 1982. "The Rise and Fall of Regulation in the Natural Gas Industry," *Public Utilities Fortnightly* (March 4) pp. 15–23.

Vernon, John M., and Daniel A. Graham. 1971. "Profitability of Monopolization by Vertical Integration," *Journal of Political Economy* vol. 79 (July/August) pp. 924–925.

Warren-Boulton, Frederic W. 1978. *Vertical Control of Markets* (Cambridge, Mass., Ballinger).

Weiss, Leonard. 1979. "The Concentration-Profits Relation and Antitrust," in Harvey Goldschmid, H. Michael Mann, and J. Fred Weston, eds. *Industrial Concentration: The New Learning* (Boston, Little, Brown).

——— and Michael W. Klass. 1981. *Case Studies in Regulation: Revolution and Reform* (Boston, Little, Brown).

Weitzman, Martin L. 1974. "Prices vs. Quantities," *Review of Economic Studies* vol. 41 (October) pp. 477–492.

———. 1980. "The 'Ratchet Principle' and Performance Incentives," *Bell Journal* vol. 11 (Spring) pp. 302–308.

Wellisz, Stanislaw. 1963. "Regulation of Natural Gas Pipeline Companies: An Economic Analysis," *Journal of Political Economy* vol. 71 (February) pp. 30–43.

Williamson, Oliver E. 1966. "Peak-Load Pricing and Optimal Capacity Under Indivisibility Constraints," *American Economic Review* vol. 56 (September) pp. 810–827.

———. 1975. *Markets and Hierarchies: Analysis and Antitrust Implications* (New York, Free Press, 1975).

———. 1979. "Transaction-Cost Economics: The Governance of Contractual Relations," *Journal of Law and Economics* vol. 22 (October) pp. 233–261.

———. 1981. "The Modern Corporation: Origins, Evolution, Attributes," *Journal of Economic Literature* vol. 19 (December) pp. 1537–1568.

Willig, Robert D., and Elizabeth Bailey. 1980. "Income Distributional Concerns in Regulatory Policy-Making." Conference Paper no. 60 (New York, National Bureau of Economic Research).

Wilson, Robert B. 1968. "The Theory of Syndicates," *Econometrica* vol. 36 (January) pp. 119–163.

Wolbert, George S., Jr. 1979. *U.S. Oil Pipelines* (Washington, D.C., American Petroleum Institute).

Wu, S. Y. 1964. "The Effect of Vertical Integration on Price and Output," *Western Economic Journal* vol. 2 (Spring) pp. 117–133.

Zajac, E. E. 1970. "A Geometric Treatment of Averch-Johnson's Behavior of the Firm Model," *American Economic Review* vol. 60 (March) pp. 117–125.

———. 1978. *Fairness or Efficiency: Introduction to Public Utility Pricing* (Cambridge, Mass., Ballinger).

INDEX

Index

About the Authors

Harry G. Broadman is a Fellow in RFF's Center for Energy Policy Research and specializes in industrial organization and regulation, and international development. He also has been associated with the Brookings Institution, the Rand Corporation, and the U.S. Energy Research and Development Administration. His work at RFF has concentrated on natural gas policy, energy problems of the third world, and issues of energy and national security. Currently he is heading up a project on oil exploration and development activity in non-OPEC developing countries.

W. David Montgomery is a Senior Fellow in RFF's Center for Energy Policy Research. Before coming to RFF, he was deputy assistant secretary in the Office of Policy and Evaluation, U.S. Department of Energy; director of the Office of Economic Analysis in the Energy Information Administration; and an analyst at the Congressional Budget Office. At RFF, he has worked on energy and national security issues, including optimal oil import levels and stockpiling policy, and on studies of natural gas markets and regulation. He is coauthor, with Douglas R. Bohi, of *Oil Prices, Energy Security, and Import Policy*.

About the Authors